The Design Engineering Aspects
of Waterflooding

Stephen C. Rose

Vice President/Manager, Production
Atlantic Richfield Indonesia

John F. Buckwalter

Vice President, Services (retired)
Ryder Scott SA

and

Robert J. Woodhall

Mechanical Engineering Assoc.
Amoco Production Co.
(deceased)

SPE Monograph Series, Volume 11

Henry L. Doherty Memorial Fund of AIME
Society of Petroleum Engineers
Richardson, TX USA

Dedication

This book is gratefully dedicated to the memory of my father,
Robert L. Rose, and my coauthor, Robert J. Woodhall.

Disclaimer

This book was prepared by members of the Society of Petroleum Engineers and their well-qualified colleagues from material published in the recognized technical literature and from their own individual experience and expertise. While the material presented is believed to be based on sound technical knowledge, neither the Society of Petroleum Engineers nor any of the authors or editors herein provide a warranty either expressed or implied in its application in the design, implementation, or analysis of waterflooding. Correspondingly, the discussion of materials, methods, or techniques that may be covered by letters patents implies no freedom to use such materials, methods, or techniques without permission through appropriate licensing. Nothing described within this book should be construed to lessen the need to apply sound engineering judgment nor to carefully apply accepted engineering practices in the design, implementation, or analysis of waterflooding.

ISBN 972-1-55563-016-4

15 16 17 18 19 / 13 12 11 10 9

Society of Petroleum Engineers
222 Palisades Creek Drive
Richardson, TX 75080-2040 USA

http://www.spe.org/store
books@spe.org
1.972.952.9393

Foreword

The authors' intent in writing this monograph was to provide a practical guide to the design of surface facilities and wells for a waterflood. The available technology and equipment in this field are constantly changing, as are the regulatory requirements. Consequently, we concentrated on comprehensive guidelines demonstrated by specific examples. The scope includes conceptual and detailed design principles, economic calculations, obtaining management and regulatory approvals, and installation. The reader will find little mathematical difficulty because waterflood design requires only basic calculations after the reservoir study is complete.

The authors have stressed practicality throughout the monograph, emphasizing the need for good engineering sense and the use of all available resources. While waterflooding is not "high tech," it is an important technique in oil recovery and has increased the world's reserves substantially. We hope that this monograph will provide immediate, useful assistance to the waterflood engineer, whether experienced in this work or not.

It is not possible to thank all the people who contributed to the writing of this monograph, but they all have my gratitude. However, I would like to acknowledge a few people individually. Without the persistence and encouragement of Joe Olds and his colleagues on the Monograph Review Committee, this work could not have been completed. Georgeann Bilich and my editor, Holly Hargadine, of the SPE staff were also a great help. I must thank my friends and coworkers at Atlantic Richfield Co. for their support and assistance. Special contributions were made by my secretaries, Linda Flynt and Ruth Dvarshkis, who remained patient and cheerful through seemingly endless drafts and rewrites, and by Garland Frasier and his drafting department of Arco Oil and Gas Co.'s Central District for the superb figures they produced. I am particularly grateful to Dick Cathriner and J.P. McDonald Jr., who expended much time and effort to teach me about waterflooding. I also owe a real debt to H.A. Koch, J.A. Middleton, and H.C. Champagne, who gave me the opportunity to work on numerous waterfloods, including Prudhoe Bay and Kuparuk. I hope that this monograph comes close to being as helpful to others as these people were to me.

Stephen C. Rose

Contents

THE DESIGN ENGINEERING ASPECTS OF WATERFLOODING

Chapter 1
Introduction

1.1 Objectives of the Monograph

This monograph is devoted to a comprehensive review of the practical aspects of designing a waterflood system. The list of disciplines involved in waterflooding includes several major branches of engineering plus such specialized areas as metallurgy, water chemistry, and corrosion engineering. This volume will concentrate on the design of surface and downhole equipment, economic evaluations of waterflood projects, and government regulation of such projects. The reservoir engineering involved in waterflooding has been handled admirably by Craig in Vol. 3 of the SPE Monograph Series, *Reservoir Engineering Aspects of Waterflooding,* and will not be covered here. Monograph Vol. 3 is highly recommended for study when any waterflood project is undertaken.

The primary objective of this monograph is to allow an engineer to plan, to design, and to provide for the installation of a successful waterflood. The reader is assumed to have been educated in an appropriate engineering discipline or to have field experience.

Waterflood engineering is defined for this monograph as the multidisciplined engineering required to plan and to install an operable waterflood. Waterflooding is the process of injecting water into an oil or, rarely, a gas reservoir to increase the hydrocarbon recovery. This is accomplished either by maintaining reservoir pressure at a higher level or by physically displacing oil with water moving through a reservoir from a wellbore dedicated to water entry to a wellbore dedicated to the removal of reservoir fluids. Waterflooding is a process system and must be handled accordingly.

A practical waterflood system consists of the application of techniques that will ensure an operable waterflood based on the experience of others without the necessity for involved mathematical calculations. This monograph will describe these techniques and their application, but the engineer must keep in mind the need for modifications to meet specific conditions. As a general rule, waterflood engineering is not extremely ''high tech'' but is very practical. Computers and other modern electronic devices are often very useful in waterflood design, but the ''garbage in, garbage out'' principle must be kept in mind. The essential word in waterflood engineering is ''operable.'' Unless the system works, the effort is unsuccessful and the project is a failure.

As mentioned previously, this monograph will not include reservoir engineering, but the waterflood engineer must work closely with the reservoir engineer to ensure that the objective of the waterflood is successfully and economically achieved. A waterflood facility may be simple, with few items of equipment, or exceedingly complex and expensive. The engineer referred to herein may be one person or a large, multidisciplined staff of hundreds. This monograph will cover the basic principles of waterflood engineering that govern the design activities of both cases.

1.2 Scope of the Monograph

This monograph will, in conjunction with the referenced documents, provide the engineer with the information needed to plan, to design, and to install an operable waterflood system. In a general sense, this process covers several phases: (1) conceptual design, (2) cost estimation and economic evaluation, (3) obtaining management and regulatory approval, and (4) detailed design and installation. The actual number of steps in the procedure, who carries them out, and the number of persons involved will depend on many factors. An independent operator may move from the oral concept to final approval in a single step. A major oil company, particularly if secondary-recovery unitization is required, may need many iterations of these steps as higher levels and more companies become involved.

To assist the waterflood engineer with both in-house efforts and dialogues with outside parties, this monograph will raise many of the issues, questions, problems, and concerns that must be satisfied before the project can proceed. There is seldom a single solution to any real-life problem. The engineer must use judgment and should not use the information in this or any other publication without critical evaluation. Even a valid principle can be misapplied in a given situation.

The monograph will emphasize planning, correct equipment selection, and project coordination. A practical approach will be used. The object is to be able to select the optimum components from the available catalog of possibilities to design an operable waterflood system. The parts of the system that will be considered include water source(s), gathering lines, water treatment, pumping equipment, injection-distribution systems, and injection wells. In conjunction with these system components, the monograph will discuss the effects of such problems as corrosion, dissolved solids, suspended solids, bacteria growth, metallurgy, and wellbore plugging.

Economic evaluations will be discussed, including cost estimation, cash flows, present worth, payouts, and similar parameters. Regulatory requirements and possible problems will be covered. Cooperative agreements and system installations are also included.

1.3 Organization of the Monograph

This introductory chapter discusses the objectives, scope, and organization of the monograph. Subsequent chapters will follow the flow path of a real project, starting with the planning effort in Chap. 2. This will cover conceptual design, feasibility, and management approval. Chap. 3 discusses the basic data collection, including geology, geography, reservoir, and production history, that leads to the basic layout design of the waterflood system. Chaps. 4 through 6 cover the design of water source and treatment systems, pump stations, injection wells, and distribution systems. The subjects included are potential water reserves, water quality measurements and requirements, processing systems, and gathering system design. The section on pump stations includes rates and pressure determinations, prime movers, pump type, and design. Injection pipeline system design is also covered, along with injection wellheads and bottomhole assemblies. Chaps. 7 and 8 deal with the changes needed in existing equipment, wells, and operations to convert from primary to waterflood operations. These include recompletions, stimulations, changing production, and artificial lift systems, consolidating batteries and gathering systems, automation, and telemetry applications. These chapters are followed in Chap. 9 by economic evaluation techniques, including cost estimation, cash flows, payout, and other calculations of economic parameters. Chap. 10 covers governing regulations, including both state and federal, common data needs, and permit requirements. Chap. 11 briefly describes such cooperative agreements as unitization and lease line injection and their effects on the waterflood engineer's work. Chap. 12 covers system installation, including specifications, purchasing, contractor relationships, quality control, and startup procedures.

This presentation is intended to be a clear, logical, and understandable one that provides a practical guide to waterflood engineering.

Chapter 2
Planning

2.1 The Project as a Problem

A project is seldom better than the effort put into the planning. This should not be taken to imply, however, that planning necessarily must be extensive or formalized. While poor planning is a serious error, overplanning is an equally grave fault. The first may lead to a failed project, but the second may lead to unnecessary delay that results in cancellation of a project. The level of planning must be appropriate to the level of complexity of the waterflood project and commensurate with the experience of the engineer or engineering team.

The first step in any planning effort is to define the problem. This certainly applies to waterflood engineering, where the project is the problem. So the first step is to define the project. Planning can then proceed to develop a solution. Muther,[1] an industrial consultant, defined the following steps in planning.

Step 1. Orientation—understand the situation.

Step 2. Prepare a solution in principle—develop an overall plan.

Step 3. Prepare a solution in detail—work out the details or specifics.

Step 4. Implementation—proceed with the project according to the plans developed in Steps 2 and 3.

Steps 1 through 3 may require a number of iterations before a final solution is achieved. Fulks,[2] another planning consultant, suggested that an operating philosophy for the life of a facility should be developed in the planning effort. Fulks also noted that planning at some point must cease and become locked in place. He suggests that design changes be put into categories of "want," "need," or "must have" and that, when time is critical, only the "must have" changes be considered.

In all planning efforts, an unknown number of iterations of the planning cycle may be required. Major changes in scope will almost invariably cause at least one iteration to be made. The size and complexity of the project, together with the number of entities involved and the type of management systems used, are major factors in determining the number of iterations needed. There may be a single pass for a small project installed by an independent producer. But a multitude of iterations may be required for a large system with several major oil company participants. The Prudhoe Bay Unit waterflood is an excellent example of the latter category, where hundreds of people and many solution iterations were required.

Planning means thinking—thinking out the concepts needed for a successful project, thinking out the tradeoffs between "good" engineering and costs. "Good" does not mean perfect engineering, which is both unobtainable and prohibitively expensive to attempt. A less-than-optimum system that is installed and operating may have more economic value than an optimal system that takes an extended time to design and to install. What constitutes "optimum" varies according to circumstances, type of operator, and time and money available. There is no absolutely correct solution to designing all or part of a waterflood system. Instead, there is an entire spectrum of good solutions, each of which may be the best for a given set of conditions. The waterflood engineer must keep this in mind and be creative and flexible to plan and to design a project.

2.2 Planning Procedures

An overall plan for project design and installation developed by an experienced project engineer is presented in Appendix A. It consists of some six phases that are analogous to Steps 1 through 4 in Muther's[1] paper. These phases are (1) development of a conceptual design, (2) development of a project cost estimate for management's approval, (3) development of a detailed design, (4) procurement of materials, (5) installation, and (6) startup. The first two phases are the basic components of planning. These will be discussed in detail in the remainder of this chapter.

2.3 Conceptual Design

The first step in conceptual design is to define or to identify the problem. Whether thought out by one man or by a task force, the seed for any waterflood will come from reservoir performance under primary depletion and, possibly, the incentive of successful waterfloods in the same or similar reservoirs. Thus, the initial impetus will come from a reservoir engineering evaluation using techniques similar to those in Craig's[3] monograph. This study will determine whether a particular reservoir will likely respond favorably to a waterflood and the probable reserves that would be added thereby. At this stage, there is probably no need for a sophisticated analysis but merely to conclude that the "stakes" are likely high enough to justify further study. With this established, the waterflood engineer will start his conceptual design work.

To develop a conceptual design, the engineer will need a certain amount of data from the reservoir engineering study. The major items needed are as follows.

1. Field name and location.

2. Producing formation, depth, and thickness.

3. Probable injection pattern and alternatives—e.g., five-spot, nine-spot, linedrive.

4. Rough order-of-magnitude estimates of injection pressures and rates.

5. Any information available on water sources and quality.

6. Any information on formation sensitivity or other pertinent items.

It is important to accept what is not on this list. A common failing at this stage is to want all the information. This is not necessary at this broad, conceptual design level. The goal here is to determine feasibility using reasonable estimates of the parameters. The end result should be the determination of whether the project has very favorable economics, unacceptable economics, or lies somewhere in between. Detailed design work to refine the results of this conceptual study will follow later if justified.

Using the information furnished by the reservoir engineering study, the waterflood engineer can lay out and size a reasonable injection-distribution system. A rough cost estimate can then be made. Horsepower requirements can be calculated from the rate and pressure data. The cost of the pump station(s) can be estimated on the basis of dollars per installed horsepower. A few assumptions and some common sense will allow a rough estimate of the cost of water source and treating systems, modifications of existing facilities, and changes in well equipment. Operating costs on a per-barrel basis can be estimated for injection and production with data from similar operations. Rough economic calculations can then be made to determine whether further work is needed (on borderline cases) or justified. It will frequently be necessary to evaluate alternative cases or solutions, which will require another iteration of the above process.

If the results of the conceptual design phase indicate that the project has a good potential for economic success, then management approval must be obtained to proceed with further study and more detailed cost estimation. This may need only a verbal approval if the project is small, particularly when the company involved is an independent producer. For more complex projects, unitization proceedings, or when major oil companies are involved, it may be advisable to produce a written report summarizing the results of the conceptual design study. A suggested outline for such a report is (1) reservoir description and geology, (2) drilling and production history, (3) waterflood results from this or similar reservoirs, (4) brief description of the water source and system design, (5) calculated waterflood results, (6) economic evaluation, and (7) conclusions and recommendations. Under some circumstances, the report on the conceptual design may be adequate for final project approval by management and no further reports will be needed.

In addition to the work outlined above, the conceptual study may also need to include permit requirements and scheduling. For larger, more complex waterfloods, it is suggested that preliminary process flow diagrams and piping and instrumentation diagrams be drawn up together with a project schedule. The level of detail and formality required will depend on the conditions governing a particular project.

2.4 Obtaining Management Approval

''Management'' as used here refers not only to in-house company management but also to outside parties such as co-owners, potential unitization partners, and their managements. The type and level of documentation required obviously depend on the circumstances. The amount of engineering effort and the expertise level dedicated is usually proportional to the size of the project and the nature of any outside participants.

2.4.1 Organization of Basic Data.

The first step in preparing a proposal for management approval is the assembly of all data available concerning the reservoir. The reservoir engineer (or group) should assemble this information, including, when available, (1) well logs and records; (2) completion data; (3) production history—oil, gas, and water; (4) core analyses; (5) bottomhole pressure surveys; (6) GOR data; (7) reservoir fluid analyses; (8) operating cost data; (9) other laboratory test data and reports; and (10) geological maps, data, and reports.

These data should then be assembled with the waterflood response predictions, waterflood system design, cost estimate, project schedule, and economic evaluation for presentation to management. An appropriate authorization for expenditure (AFE) form should be prepared for approval. It is probable that further refinement of the cost estimate from the conceptual study will be needed because that estimate is generally no better than $\pm 20\%$. Under certain conditions—for example, if time is critical and the project has very favorable economics—this rough estimate plus 10 to 20% contingencies can be used for AFE approval. In most cases, however, the required accuracy for AFE approval is in the ± 5 to 10% range.

Further definition of chosen alternatives and iterations of the cost estimate will lead to an AFE quality estimate. At this point, the engineering effort becomes more specific, with alternatives evaluated and choices made. It is very helpful to obtain information from manufacturers' representatives. These representatives are usually willing to spend a reasonable amount of effort showing how their product(s) would apply to the case at hand. Generally, most manufacturers will be glad to provide an engineering cost estimate for their product in a specific application. However, the waterflood engineer must not create the impression that a specific piece of equipment will be purchased from a given supplier. At this point in the planning, only information is being sought. Manufacturers' representatives may also be helpful in writing specifications for equipment.

Engineering estimates cost the vendors money to develop, which is eventually passed on to the purchasers. The engineer should not request estimates from vendors who would not be qualified bidders when the equipment is purchased. Nor should the engineer ask for estimates on an excessive number of alternatives or on too tight a set of specifications. The engineer should also be very careful not to let the vendors know (even approximately) what their competitors have given as estimates. In most companies, the waterflood engineer is not the final purchaser. The purchasing contracts are usually completed by a buyer in the company's purchasing department. Getting engineering estimates as contrasted to quotations will usually maintain good relations between the waterflood engineer and the purchasing agent. These buyers are very knowledgeable about current pricing and can be a valuable resource to the engineer during the process of estimating project cost. After going through the above procedures, the waterflood engineer should have an AFE quality estimate and an economic evaluation based on that estimate. If the working interest in the project is entirely owned by the company, the engineer is now ready to try for management approval.

In cases where unitization will be necessary, a number of additional items are usually required. These may include isopach maps, cross sections, and isoporosity maps. Other data that frequently enter into unitization proceedings are surface area, number of completions, current and cumulative production, gross pay, net pay, original oil in place, remaining primary reserves, and estimated secondary reserves. While most companies use their reservoir engineers for this type of effort, the waterflood engineer should be aware of these needs and be prepared to include them in presentations or reports as appropriate.

2.4.2 Management Approval Procedures.

Once the necessary data have been gathered and organized into a report, the engineer must present this for management approval. Major companies usually have procedures that differ considerably from those of independent producers; therefore, both will be discussed here. The specific paperwork and procedures will obviously vary from company to company, so only general principles will be covered.

Each company will set some minimum economic criteria that must be met by its investment projects. The obvious minimum is that probable income must exceed expected outgo when the cash flow is reduced to a present-worth basis. Further, the present-worth discount factor should exceed the cost of money to the company. For example, if the company is able to borrow money at 10% and the undiscounted present worth of the project is 10% after taxes, the company has paid the interest and no more. Stated differently, if a given project has a present worth of zero calculated at a 10% discount factor, the company would do as well to deposit the money in a bank at 10% interest. Company management may still approve the project because of the desire to stay in business with a 10% rate of return.

It is expected, then, that each company will set a criterion that the rate of return be above its rate to borrow money. Further, the company probably evaluates its spending on the basis of a ranking of the return from various projects; i.e., a project with a calculated return on invested money of 50%/year would have a much higher priority than one with a return of 10%/year. This all presumes that there are a list of projects to choose from and no extenuating circumstance involved. Major companies will normally consider or include in their calculations some measure of the risk of failure.

The specific criteria of his company will be known to the waterflood engineer and will be calculated and included in reports or presentations to management. In a major oil company, at least two or more levels of approval will likely be needed. The size of the investment usually determines the number of levels that must be traversed for final approval.

Each level has its own requirements for completeness of the backup documentation. It is probable that the lowest levels will require the greatest amount of technical input. As the corporate ladder is climbed, the emphasis will change to monetary interests.

The lowest level will be involved in deciding to make a more detailed evaluation and in subsequent possible preparation of a proposed budget item. Middle management would normally approve the start of unitization proceedings as well as the inclusion of the project in the budget (on the basis of the conceptual study). Middle management would also authorize expenditures for a medium-cost project, the final design, the right to request bids, and the initiation of project work. Upper management will probably be limited to very expensive projects requiring this level of authorization. Its approval will be more dependent on economics than technical input. Upper management will assume that the technical aspects have been appropriately covered at the lower levels.

In a smaller company, the entire process can be much simpler. The relationship between the waterflood engineer and the manager with authority for final approval may be one-on-one. The circumstances will generally be very informal, with little or no involvement of committees, task forces, or long, formal reports. While the same basic kinds of data, cost estimates, economic evaluations, etc., are needed, very few layers of management (frequently only one) need to be satisfied. The waterflood engineer will commonly be more dependent on outside help, such as engineering consultants and manufacturing representatives, to obtain the information required. Given that these differences exist between types of companies, it must be emphasized that the basic planning process the waterflood engineer goes through is the same. The differences

should not be allowed to obscure the fact that good planning is essential to obtaining management approval and ensuring that a successful project results.

There is *no* substitute for thinking out the best solution to the problem. This is planning at its simplest level. Mistakes should be made on paper where an eraser can remove them, not in the field where someone must live with them.

References

1. Muther, R.: "A Planner's Planner Reveals the Secret to Greater Productivity," *Engineer's Digest* (May 1982) 8-9.
2. Fulks, B.D.: "Planning and Organizing for Less Troublesome Plant Startups," *Chem. Eng.* (Sept. 6, 1982) 96-106.
3. Craig, F.F. Jr.: *Reservoir Engineering Aspects of Waterflooding,* Monograph Series, SPE, Richardson, TX (1971) 3.

Chapter 3
Information and Data Analysis

3.1 Field Information

As discussed in Chap. 2, the first step in developing a waterflood is to define the reservoir and to estimate the results of waterflooding. This task is generally assigned to a reservoir engineer or a reservoir engineering group, which may be in-house or an outside consultant. If a unit is to be formed, this person or group will usually be assigned to a technical committee that will study and define the reservoir and waterflood results jointly.

An evaluation of this type includes consideration of the physical characteristics of the reservoir, such as permeability, porosity, thickness, areal extent, oil/water/gas interfaces, wettability, and oil/water/gas properties (phase behavior). During the study, other factors will also be examined, including the geologic history of the formation, any reservoir anisotropy such as oriented fracture systems or directional permeability, unusual stimulation techniques used, production history, physical layout of the wells, topography, alternative injection patterns, and other items that may be specific to the site or reservoir.

The results of this study will provide the waterflood engineer with recommendations on injection patterns, injection-well selection (whether new wells or conversions), injection rates and pressures, potential water sources, and other information needed to start the design process. A recommendation will usually be provided concerning whether a pilot test, staged waterflood, or fieldwide installation should be initiated. The remainder of this chapter discusses some of these areas in more detail.

3.2 Basic Geologic and Engineering Data

3.2.1 Physical Boundaries. The first step in defining the reservoir is to determine its areal and vertical extent. The edge of the productive limits must be located, and the means by which these limits are defined must be determined. For example, if the edge is defined by a water/oil contact in the reservoir, then future decisions may be much different from those for a reservoir defined by low permeability and a lack of productivity. The former reservoir may be suitable for injection into or near the water zone around the edges. The latter type would have little or no injectivity in the edge wells, suggesting the probability that oil will be trapped against the permeability pinchout and not be recovered. Hence, the chosen pattern for waterflooding would be different for these two types of reservoir limits. It is important to define both the areal and vertical limits of the zone of interest. Plans will vary considerably depending on the factors that determine these limits, such as gas or water interfaces, permeability pinchouts, and partially or wholly sealing shales or faults.

The areal configuration of the project is very important. A "shoestring"-type reservoir, for example, requires a well-pattern arrangement quite different from a project in the extensive San Andres dolomite of west Texas. As the name indicates, the former type has a reservoir configuration that is long and narrow and would most likely be developed on an end-to-end type of flood or by use of one of the linedrive patterns. The west Texas dolomite generally contains very large reservoirs, covering several square miles of surface area. As discussed below, these reservoirs are usually best developed by one of the repeating pattern types such as the five-spot.

If the formation to be waterflooded is fairly small in area and at least vertically homogeneous, a peripheral waterflood may be used profitably, especially if the wells have good injectivity. The peripheral flood requires that wells be drilled for or converted to injection on the edges of the reservoir. Unfortunately, it is common for wells on the edge of a reservoir to be of poor quality. The pay zone may be discontinuous, thin, or with poor porosity and low permeability. The peripheral flood must sweep the reservoir as a whole, so the rate of water injection into the edge wells must be high enough to achieve fill-up in the entire reservoir within 1 to 2 years. Longer fill-up times are generally uneconomical. A number of peripheral floods in larger reservoirs have shown poor results until they have been converted to pattern floods. An example of this is the Orcutt flood in California.[1]

3.2.2 Physical Reservoir Characteristics. Many physical characteristics of the reservoir rock and fluids can have significant influence on the success of a waterflood. Some of these factors are obvious, such as porosity, permeability, bulk volume, oil and gas saturations, and the phase behavior and viscosity of the reservoir fluids. The effect of these factors has been more than adequately handled in Craig's[2] monograph and other publications. However, other characteristics, such as reservoir heterogeneity, are important to waterflood design. This section discusses some of these factors that merit consideration.

3.2.2.1 Pay Quality and Continuity. Before waterflooding technology developed to its current level, some people postulated that a reservoir that produced no more than one-half the expected normal recovery per acre-foot during full primary depletion would be a good waterflood prospect because more than the usual amount of oil remained in the reservoir. In many cases, however, poor recovery is associated with poor-quality pay, which will also cause poor waterflood results.

A successful waterflood requires some continuity from injection to production wells in all zones or leases to be flooded. To flood a zone with low permeability or porosity and with poor continuity from well to well profitably, a closely spaced injection pattern will be required with a ratio of at least one injection well to one producing well, as in the five-spot.

On the other end of the scale, high-quality pay that is continuous over the majority of the reservoir can be flooded economically by virtually any plan desired, although some may be better than others. Good-quality, continuous pay means a high-porosity and high-permeability formation in which oil may be moved by water over large distances as the result of good communication across the zones. Linedrive or peripheral-type flooding is possible in these reservoirs and may be more profitable than a denser pattern development.

3.2.2.2 Zonation and Heterogeneous Effects. In general, oil reservoirs are characterized by alternate layers of permeable and impermeable rock. In the case of sandstone reservoirs, these layers are sandstone and (usually) shale. The permeable layers are called pay zones. A log or core analysis report for a well may show that the well has a certain number of pay zones. Each zone will have different characteristics, such as permeability, porosity, thickness, and fluid saturations. Often it is assumed that the individual zones are completely isolated from each other and that the only possible vertical flow of fluids occurs at the wellbores. If this is true, then each zone is an individual reservoir.

Some sand zones have very low vertical permeability or they may contain thin layers of impermeable material, such as mica plates, so that the zone may have several internal layers. These factors may prevent effective movement of fluids between the subzones.

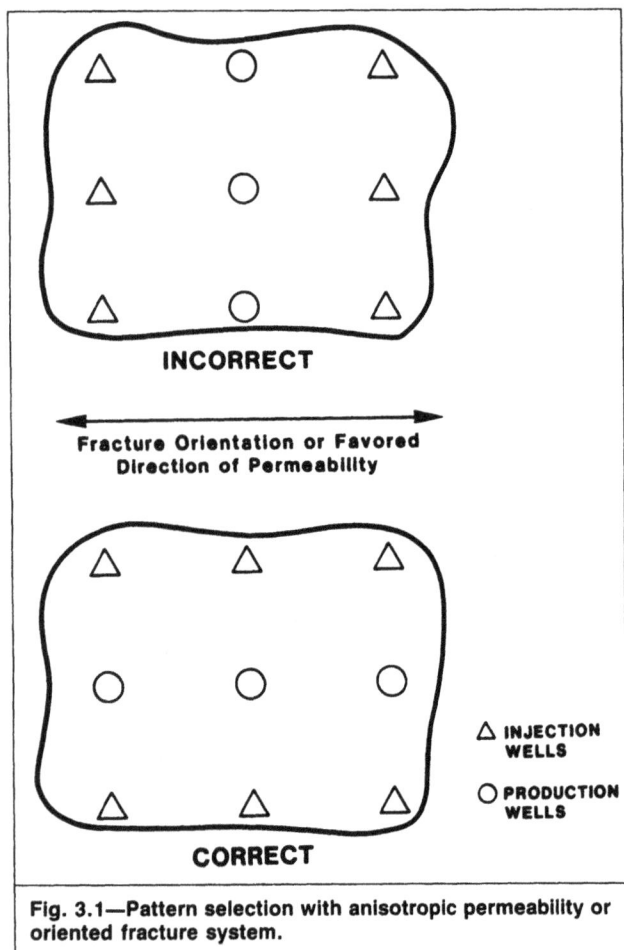

Fig. 3.1—Pattern selection with anisotropic permeability or oriented fracture system.

In such cases, each subzone is a separate oil reservoir. When waterflooding is initiated in reservoirs of this type, the most permeable zone or zones will flood out first. This generally results in an irregular flood front at most producing wells. Some zones may be producing oil, some water, and others may not be filled up. All these stages may occur simultaneously in the same well.

The technique of shutting in or converting wells as water cut increases is common in peripheral floods but does not work well with a reservoir consisting of a number of isolated zones. One of the denser patterns, such as a five-spot, is a better choice with zoned reservoirs. The producing wells in five-spots will continue to operate throughout the flood life, which allows each of the zones to produce its oil in turn, depending on its permeability. In addition, the five-spot pattern generates a higher pressure gradient, which helps to sweep the low-permeability zones more effectively than peripheral flooding.

Producing wells must be kept pumped off with a minimum fluid level, thus lifting fluids as they are available at the wellbore. A high fluid level will impose a backpressure. In such cases, zones that are not pressured adequately may not produce, or crossflow between zones may occur. This can delay or prevent oil production from the less-permeable sands and result in an economic loss.

Other reservoirs may appear in individual wells to be made up of independent pay zones. Between the wells, however, the pay zones can have good vertical permeability, so that there is communication between the sand layers. If this interlayer flow allows all zones to be swept properly by the water, these reservoirs will behave similarly to the continuous-pay-zone reservoirs discussed in Sec. 3.2.2.1. The entire reservoir will reach fill-up at about the same time and the vertical flood front will move forward fairly evenly. Some reservoirs have no crossflow between layers and others have very high vertical permeability between layers. Most fields lie between these two extremes.

Areal heterogeneity of oil reservoirs seems to be most common. It is a very exceptional reservoir that has little variation in permeability as measured from well to well. Even with closely spaced wells

(as in the Bradford field in Pennsylvania, with 1 to 4 acres/well [0.4 to 1.6 ha/well]), the areal variation in permeability is significant when considered on a well-to-well basis or within any individual pay zone. Heterogeneities are not limited to permeability; reservoir thickness varies, as does porosity. Oil and gas saturations may vary to a marked degree, depending on the production history and secondary recovery attempts in the field.

If the reservoir is highly heterogeneous, repeating-pattern arrangements of injection wells tend to be superior to linedrive or peripheral flooding. One effect of the dense, repetitive injection patterns is to force fluids to flow more directly over the limited distances from injection well to producing well. This control exercised on fluid flow generally produces improved sweep efficiency compared with allowing fluids to move through the reservoir along paths of least resistance. The latter flow system causes the flood front to become very uneven and may allow oil to be bypassed. Again, the more homogeneous reservoirs may be flooded by peripheral methods. The more heterogeneous ones tend to respond best to repeated intensive-pattern flooding.

3.2.2.3 Permeability Direction. There are pay zones where the permeability is greater in one direction than in other directions. The zone, or formation, may not be known to have this directional permeability characteristic before water is injected into it.

A long, narrow reservoir, such as a shoestring type, may be suspected to have more favorable permeability in the direction of its longest dimension. It is frequently best to drill linedrive injection wells on close spacing across the narrow dimension. The distance between a row of injectors and a row of producers will be two or three times the spacing between injectors in the row. This is one method of compensating for anisotropic permeability in this type of reservoir. As a general rule, however, where directional permeability exists, injection wells should be selected so that a line drawn through adjacent injectors parallels the highest-permeability trend as closely as possible. Fig. 3.1 demonstrates one example of this type of pattern.

Some oil reservoirs have high-permeability streaks or areas, sometimes called "sweet spots." Wells that penetrate and produce from these sections have abnormally high rates and reserves compared with average wells in the field during primary depletion. The initial rates and subsequent cumulative production of the sweet-spot wells may be on the order of two or three times as high as an average well in the same reservoir. It may be difficult to design a pattern to take these areas of favorable permeability into consideration. If these excellent areas are in a distinct trend, it would be advisable to line up injectors parallel to the trend for improved sweep efficiency.

Pulse testing[3,4] is a procedure that measures reservoir characteristics between wells, including permeability. This procedure requires a series of flow disturbances generated at one well while pressure readings are recorded in another well. Alternate periods of flow and shut-in (or injection and shut-in) create the disturbances that are measured. This technique may be used to detect areal anomalies in permeability as well as directional permeability trends. Other procedures, such as tracer testing and careful study of the production history, can also be helpful in analyzing anisotropic or areal variations of permeability.

3.2.2.4 Natural Fracture Orientation. Natural fractures of the pay formation that are long enough to allow direct fracture flow of injected water from injector to oil producer cannot be ignored if the reservoir is to be waterflooded successfully. Many large oil reservoirs, including sands, carbonates, and shales, have natural vertical fractures or joints. In most instances, these fractures are unidirectional and parallel.

If a line between injection wells and producing wells is in the same direction as the vertical fracture, a waterflood failure results because of premature water breakthrough. The best pattern arrangement is one in which a line drawn between adjacent injectors is parallel to the fractures (Fig. 3.1). Successful floods exist in reservoirs with many large-scale fractures where injection wells located parallel to the fractures are used.

Some pay zones have incipient vertical fractures that open up with injection of water under pressure. This problem has the same solution as stated above; i.e., rows of injectors must be approximately parallel to the fractures.

Natural horizontal fractures are found rarely, if at all, in oil reservoirs. It is possible to create a horizontal fracture by injecting water at very high pressure in shallow formations. Overburden lifting has occurred by high-pressure injection. If a horizontal fracture is induced at the injection well and, as it extends, it then intersects a plane of weakness or a fracture in the vertical direction, the fracture will change to and continue in the vertical orientation. Because most or all natural fractures are vertical, it follows that most induced fractures are or become vertical. Normally, injection pressures at the face of the formation should be restricted below the level at which fractures are initiated.

Naturally occurring vertical fractures are the ones that must be dealt with in waterflood design. It is fortunate that these fractures are usually parallel to each other throughout a reservoir and can be handled through selection of the proper injection pattern.

3.2.2.5 Unusual Completions. In certain fields, the completion procedures in some or all wells may lead to difficulties in waterflood design. Completions that prevent adequate control of the injection or production of fluids in wells are of particular concern. Openhole completions that include the entire pay interval may prevent control of injection by zone or restriction of injection to certain zones. Such a completion may make it impossible to eliminate injection into "thief" zones (high-permeability zones). Similar problems arise in shutting off high-water-cut zones in openhole producing wells. Other types of completions that can create these problems are those that have been fractured or blanket-perforated over the entire pay interval, particularly with a high number of perforations per foot.

Wells completed in more than one reservoir may also cause significant problems when waterflooding is initiated. Mechanical failures or difficulty in using such wells in the waterflood may result from these completion techniques. Fractured wells, particularly those with large-volume treatments, may not be suitable for conversion to injection because the fracture can have a significant effect on reservoir sweep efficiency. If it is necessary to use fractured wells, they can sometimes be handled with the same pattern choices as for naturally fractured reservoirs because the induced fractures tend to have the same orientation.

Unusual completion techniques in a field do not preclude a successful waterflood, but they must be evaluated thoroughly. Mitigation of the effects is often very possible with various profile-control techniques, recompletions that change the mechanical condition of the well (liners and sliding sleeves), injection-pressure control, and pattern selection. The key element is recognition of the potential problem on the basis of field data and then development of the appropriate economic solution.

3.3 Primary Operations History

Three areas of the field operations history of particular importance in waterflood design are discussed in this section: the production history under primary development, production equipment installed, and well-completion data.

3.3.1 Primary Recovery Data. These data are perhaps more important to the reservoir engineer but can be very useful to the waterflood designer. Analysis of the production history by well and zone can help in detection of directional permeability trends, fracture systems, and areas of favorable permeability. While many other factors affect the production from any one well, analysis of field trends may expose some of these potential problem areas. The distribution of oil and gas saturations can also be estimated on the basis of the production history. The waterflood engineer is well advised to evaluate the performance during primary depletion and to review this with the reservoir engineers. For example, the waterflood engineer may notice particularly high cumulative production per well in a certain part of a field. Discussions with the reservoir engineers may show this area to have very favorable permeability and porosity. Another possibility may be that additional reservoir thickness or even another pay zone exists in that area. Areas of low primary recovery may denote low reservoir pressure but frequently are a result of poor formation quality. Whatever the case, discussions with the reservoir engineers regarding the anomalies and characterization of the primary production phase will assist the waterflood engineer in designing the facilities for a successful project.

3.3.2 Production Equipment Installed. The size and condition of production equipment are very important in estimating costs for a waterflood. A number of waterfloods have been economic failures because of additional expenses over the initial estimate necessitated by replacing the existing production equipment.[5] During the life of a waterflood, the producing wells will make more water and usually more total fluid than during primary depletion. The pipelines installed for primary production are frequently undersized for waterflooding. Conversely, the facilities that gather, treat, and compress gas may be oversized.

A careful evaluation of the existing equipment is needed to ensure that it will handle the expected waterflood production. If it will not, then the cost of replacing the equipment must be included in the waterflood economics. In most cases, it will be necessary to upgrade the fluid-handling systems and the artificial-lift equipment.

The condition of the existing facilities is also important. By the end of primary depletion, the equipment may have deteriorated from corrosion, poor maintenance, or other causes. This can be a particular problem when a waterflood unit is proposed. The waterflood engineer should make a field inspection to evaluate the condition of all the facilities to be included in the project.

3.3.3 Well-Completion Data. The type of completion and stimulation history was discussed in Sec. 3.2.2.5. It is also essential to examine the downhole wellbore conditions from a mechanical standpoint. The condition of the casing can be critical, particularly in wells that are to be converted to injection service. Federal and state regulations require that casing integrity be proven before injection is started. This test usually consists of pressuring up the casing/tubing annulus to the maximum allowed injection pressure or at least 300 psig [2068 kPa]. Because failing this test may result in expensive repair work or even drilling a new well, the condition of casing in the existing wells can have a significant impact on waterflood costs.

Cement tops and cement bond logs (where available) should be evaluated when planning a waterflood. Poor cement bonding in the interval to be flooded may allow water to flow from or into zones when that is not desirable. If the bond is exceptionally bad, water could flow behind the casing, possibly contaminating freshwater zones. The size and condition of the tubing in the wells also should be investigated. In injection wells, the tubing should be of adequate size and condition to allow injection at the desired rates and pressures without excessive pressure drops or leaks. The tubing should also be large enough to allow passage of injection logging tools. In producing wells, the tubing should be sized to allow the expected rates to be lifted. Waterflooding will often cause increased corrosion and scale problems, particularly in production wells. The tubing should be sound enough to prevent these problems from becoming critical.

Evaluation of the mechanical completion of the wells may also include consideration of any special hardware installed, such as sliding sleeves, downhole chokes, packers, anchors, tubing stops, or other items that may affect the use of existing wells in the waterflood. Careful attention by the waterflood engineer is essential to prevent serious operational problems in the future.

3.4 Waterflood Layout

Craig[2,6] discusses different waterflood patterns, emphasizing that the pattern chosen should (1) provide the desired oil production capacity; (2) provide sufficient injection capacity to support the oil production; (3) maximize oil recovery with a minimum of water production; (4) use reservoir heterogeneity to the best advantage; and (5) use the existing wells to minimize the number of new wells required.

A reservoir study is necessary for a proper choice of pattern selection. As mentioned earlier, this type of study is usually performed by in-house or consulting reservoir engineers. The pattern should be chosen on the basis of the first four characteristics mentioned by Craig. The chosen pattern may then be modified to conform with the fifth criterion.

3.4.1 Potential Pattern Selection. The five-spot waterflood pattern has been used more frequently than any other. Second to the five-spot in popularity is the peripheral or linedrive flood pattern,

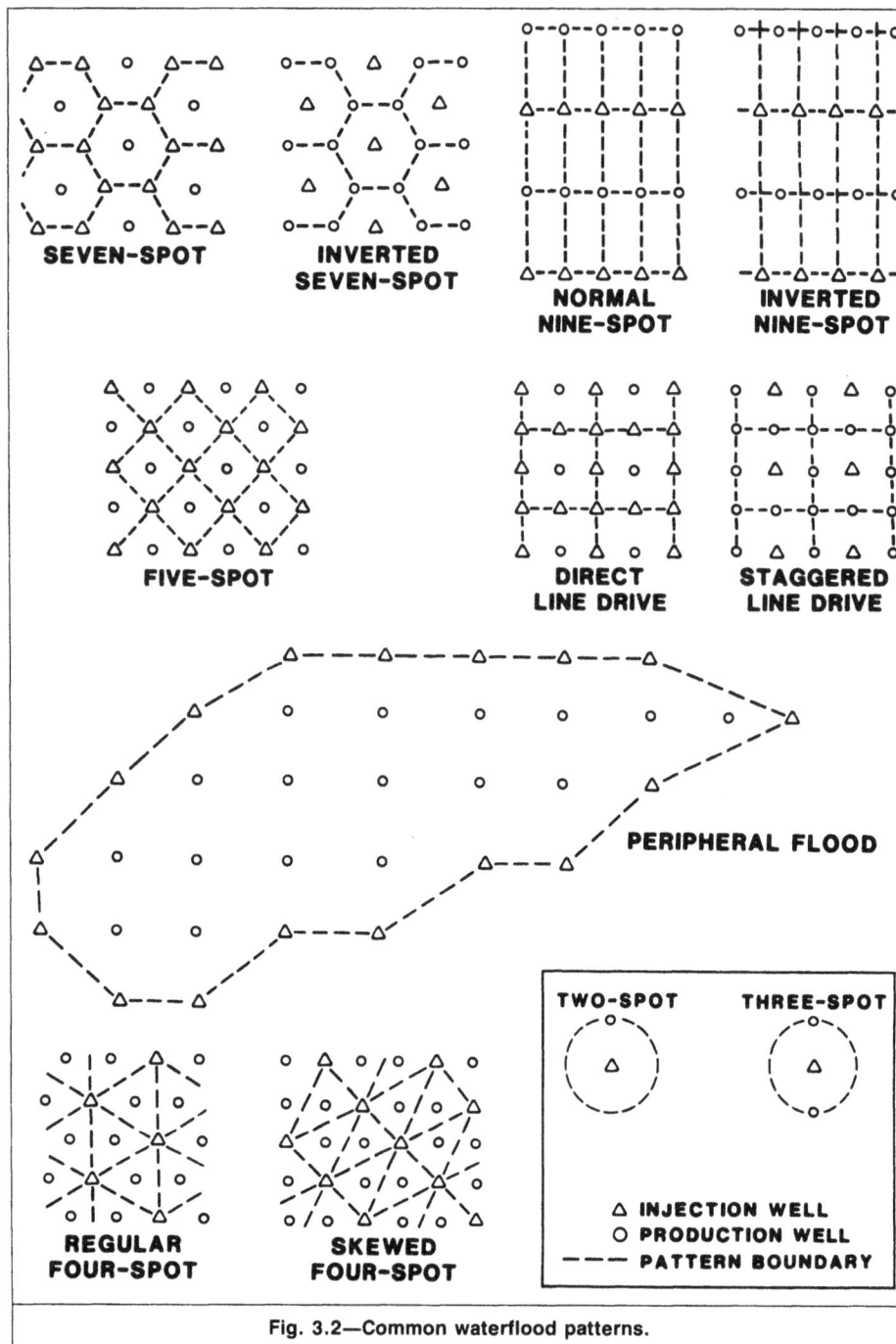

Fig. 3.2—Common waterflood patterns.

and in third place is the inverted nine-spot. Many other patterns have been tried less frequently. One reason for the popularity of the five-spot is the good results it has shown. Because of well-spacing regulations, primary wells are usually drilled on a square pattern, which lends itself easily to conversion to a five-spot waterflood.

When a flooding pattern is chosen, the engineering decision should be made on the basis of the reservoir study and the factors discussed in this chapter. One way to proceed would be to evaluate the predicted injectivity per well, the mobility ratio, and the degree of heterogeneity of the reservoir. If the injectivity is low or the heterogeneity is large, a five-spot may be the right choice. If the mobility ratio is unfavorable, an inverted nine-spot should be considered, provided that the reservoir is reasonably homogeneous. If the injectivities are good and the reservoir is not too large, the peripheral flood, linedrive, or a similar type should be considered. Anisotropic permeability, permeability trends, or oriented fracture systems may also favor peripheral or linedrive patterns. Fig. 3.2 shows some of the more common patterns. Numerous published

reports describe how various patterns were used in or modified for specific fields. One good example is the waterflood in the Norge Marchand Unit, developed as follows:

"Plans called for total water injection rate of 35,000 b/d through 23 input wells; 21 converted producers; and, two newly drilled wells. An irregular injection pattern was developed to conform to reservoir geology. In the more permeable, thinner sands in the north half of the field, injection wells were selected at alternating locations along the channel deposit. In tighter, thicker deposits to the south, injection wells were more densely concentrated in a five-spot pattern, somewhat skewed due to irregular location of wells within each 160-acre drilling and spacing unit."[7]

As can be seen in this example, the patterns selected were substantially altered to fit the circumstances in the Norge Marchand Unit. No two oil reservoirs are exactly alike. Where reservoirs are similar, experience gained in one waterflood may be applicable to another. Flood-pattern selection is difficult because of the large variety of reservoir types, but prior experience can be used to help make this decision.

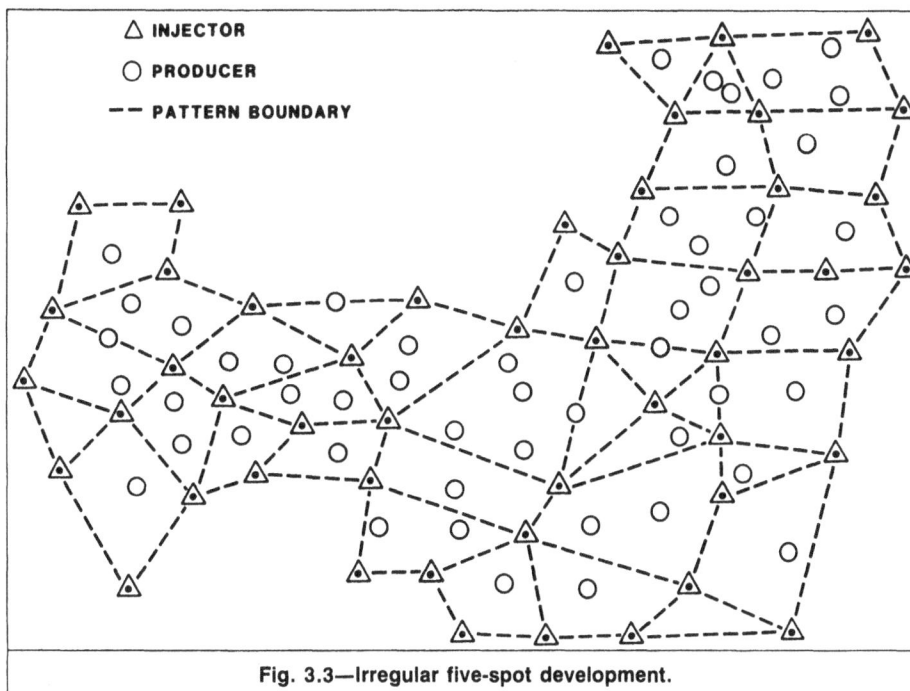

Fig. 3.3—Irregular five-spot development.

3.4.2 Selection of Wells and Spacing. Selection of wells for injection and production is based on the pattern chosen for development. This should be modified as needed, however, to fit the conditions of the specific field. The existing well pattern may be highly irregular and may tempt the engineer to abandon the use of a regular pattern such as a five-spot. This need not be done. For example, the five-spot arrangement can be used as a basic guide and the wells selected to approximate a 1 : 1 ratio of producers to injectors. Fig. 3.3 shows an example of an irregular five-spot pattern that has produced an exceptionally good response to the flood. Frequently, there is a temptation to adjust a pattern to select the poorest producers as injectors to reduce the loss of production. The waterflood engineer should not yield to this argument. Poor producers make poor injectors and reduce or delay the waterflood response.

Well spacing for flooding is usually determined by the primary drilling spacing. The advantages of using the primary spacing are that drilling new wells becomes unnecessary and production equipment is already available for these wells. If the primary spacing is too wide, drilling new infill injection wells will solve the problem where economically justified.

In the past, some older waterflood areas in Pennsylvania, Ohio, Illinois, Kansas, and Oklahoma required completely new drilling to initiate water-injection projects. In the oldest fields, primary production had ceased. Frequently, the casing had been salvaged and the production equipment removed. The operators, usually by trial and error, arrived at satisfactory well spacing for individual reservoirs. They then proceeded to drill and to waterflood, usually on the five-spot pattern. Corrections in spacing of wells were made in newly developed properties by evaluation of results at neighboring properties. Today, the most likely correction to primary spacing will be the addition of infill wells where pay formations are of low permeability and are very heterogeneous.

3.4.3 Production Loss From Injection Wells. Production losses resulting from conversion of producing wells to injection wells are losses of current production only. This could be called a rate deceleration case. The rate of oil production for the property or unit is reduced by conversions. With a successful waterflood, the oil production rate will be restored and enhanced.

The loss of current production can be a serious matter, especially for operators of small tracts in a unit. In the extreme case, an individual may own only one oil well producing, for example, a few barrels of oil per day. To lose all this production because of the conversion of the well to injection may be a financial catastrophe for the individual. On the other hand, a major oil company

will find this temporary loss of income from one small well inconsequential.

Waterflooding often is initiated after the reservoir pressure is very low and the reservoir has produced nearly all its primary oil reserves. In some instances, there is no current production on the property when waterflooding is initiated. In many fields, new wells have been drilled specifically for the purpose of waterflooding. Sometimes, the injection wells are all new infill wells drilled for that purpose. In these cases, there is no loss of current production by conversion of wells to injection. On the other hand, many waterfloods are initiated during the primary life by conversion of wells to injectors. Here the loss of current production may range from as much as one-half of the total in a five-spot pattern to one-fourth or less in a peripheral development. Because of the individual loss of current income by use of the five-spot pattern, there may be pressure to use a peripheral approach in a reservoir not suited for that type of development.

At times, justification is presented for use of an inverted nine-spot pattern because the ratio of producers to injectors is 3 : 1, which reduces loss of current income compared with the 1 : 1 ratio of producers to injectors in the five-spot pattern. Choosing the nine-spot may be an excellent decision if the reservoir is suitable; i.e., it has exceptionally good injection capacity per well, has good sand continuity, and is reasonably homogeneous. However, if the reservoir requires five-spot development because of low injection capacity per well or because of reservoir heterogeneities, using the nine-spot is a serious error. Either of these examples demonstrates poor engineering judgment. The economic loss from a project developed by an incorrectly chosen pattern is usually much more than the temporary loss of current income caused by conversion of producers to injectors.

When a water-injection project is started early in the primary producing life of a reservoir (which is generally the best time to start water injection), the loss of current production because of conversions lasts only a short time because the delay required to fill up the reservoir is minimized. Several major fields, including Prudhoe Bay and Kuparuk on the North Slope of Alaska and Ninian in the North Sea, have initiated waterfloods early in their primary development periods.

3.4.4 Possible Expansion. There may be reasons for staging the development of some waterflood properties or units. This may cause a considerable time lag between water injection in the first stage and subsequent water injection in the second stage. If, for example, only one-half of the property has injectors taking water for 1 year or more in an overall pattern development, the undeveloped half will be partially flooded by injectors at the edge of the first-stage development. A seriously unbalanced pattern of fluid flow into the first row of producers of the unflooded half of the property will then occur. This unbalanced flow cannot be avoided under a two-stage development plan with a significant time delay between stages. It is possible that some oil reserves will be lost in the process. This is the price that must be paid for the convenience of staging. Delay should be avoided unless there is an economic advantage to staging that will compensate for any oil losses.

In most instances, Stage 2 will be initiated by expanding the Stage 1 flood to adjacent areas using the same pattern. An adjacent row of producers will be converted to injection as required to fit the pattern symmetry. The remaining oil producers will not respond normally to the new set of injectors. They may have to produce at high water cuts for a longer-than-normal period. The second row of producers may produce some oil that has bypassed the first row of producers.

Where reservoir conditions permit, a peripheral pattern may be used, in which there is normally a long lag between the conversion of one set of wells and the conversion of the next set to water injection. In this case, there may be no losses of oil by delayed development. Unfortunately, however, most reservoirs are heterogeneous and some oil loss is likely to result from long delays in the development of a flood.

3.4.5 Terrain and Topography.
Terrain and topography may have a significant effect on the choice of pattern or well spacing. Unusual terrain such as marshes, offshore environments, urban developments, tundra, unstable soils, and rivers, ponds, and other bodies of water may require the waterflood engineer to modify the pattern and choice of wells. His equipment choices may also be restricted by these factors. The environment should be carefully investigated by the engineer to determine the conditions that exist and how best to mitigate their effect.

Similarly, severe topographical conditions may place restraints on the waterflood design and will certainly affect costs. A field located in the mountains will require much more design work than one on the plains of west Texas. The effects of elevation changes on the pressures and flow conditions in gathering and injection systems can create substantial piping design problems. Rough terrain may require the use of helicopters to lay lines, which significantly increases costs.

The potential problems associated with these conditions are sufficiently complex and so site-specific that they cannot be dealt with in detail in a monograph of this kind. However, the waterflood engineer should ensure that these factors are thoroughly examined and taken into account in the waterflood design.

3.5 Pilot Tests
Any prediction of reservoir behavior under a waterflood has a high level of uncertainty. As Craig[2] states, "the oil reservoir is a very complex system of solids, liquids, and gases. The reservoir engineer must make predictions of the behavior of this complexity when water is injected into it and accurate prediction is a difficult problem." The inadequacy of available data causes much of the difficulty in waterflood predictions. Engineers frequently make calculations of reservoir response that turn out to be very different from the actual waterflood behavior.

As a result of the uncertainty associated with waterflood performance predictions, methods of obtaining field data to verify the response calculations are very helpful. Some of these have been discussed earlier in this chapter. One useful tool is the pilot waterflood project. Normally, this will involve choosing a typical area of the reservoir and installing a waterflood. If the field is expected to respond best to a pattern flood, then one or two individual patterns may be used for the pilot test. It is critical to choose an area that will be characteristic of the entire reservoir. Predictions for this specific section can be made and then compared with the actual performance. The results of this comparison can be used to calibrate a simulation model, extrapolated to full-field performance, or used in other ways to verify the reservoir prediction.

A pilot test program will cause a delay in the implementation of a full-field waterflood. The economic disadvantage of the delay may be offset by the more accurate prediction of waterflood response. In some cases, the pilot-test results prevent the installation of a full-field waterflood that would fail. Normally, the economics will favor a pilot test whenever the field is fairly large and the reservoir prediction has a high level of uncertainty. The waterflood engineer should evaluate the performance prediction carefully before recommending a pilot test. The economics and risk associated with the fieldwide project will determine the need for a pilot waterflood test.

References

1. Stracke, K.J., Case, C.A., and Baer, P.J. Jr.: "Orcutt Third Zone Waterflood: A Successful Project," paper SPE 1542 presented at the 1966 SPE Annual Meeting, Dallas, TX, Oct. 2–5.
2. Craig, F.F. Jr.: *The Reservoir Engineering Aspects of Waterflooding*, Monograph Series, SPE, Richardson, TX (1971) **3**.
3. Johnson, C.R., Greenhorn, R.A., and Woods, E.G.: "Pulse-Testing: A New Method for Describing Reservoir Flow Properties Between Wells," *JPT* (Dec. 1966) 1599–1604; *Trans.*, AIME, **237**.
4. Kamal, M. and Brigham, W. E.: "Design and Analysis of Pulse Tests with Unequal Pulse and Shut-In Periods," *JPT* (Feb. 1976) 205–12.
5. Jackson, R.W.: "Why Some Waterfloods Fail," *World Oil* (March 1968) 65–68.
6. Craig, F.F. Jr.: "Engineering Waterfloods for Improved Oil Recovery," *Pet. Eng.* (Dec. 1973) 23–30.
7. Jowell, Jack H. and Haberthur, Charles R.: "Norge Marchand Deep Injection Unit," *Pet. Eng.* (Nov. 1976) 23–33.

Chapter 4
Water Sources and Water Treatment

4.1 Introduction

This chapter considers water sources and their potential problems. Treatment processes needed to produce acceptable water quality are also discussed. A lack of available water could prevent effective waterflooding, and poor quality water may result in a waterflood failure. Improper or inadequate water treatment can also lead to problems that cause economic failure. The subjects covered in this chapter are among the most critical to be considered in waterflood design.

The ideal water for waterflooding would have the following characteristics: (1) no suspended solid particles, (2) no harmful dissolved solids causing scaling or corrosion, (3) no dissolved gases, (4) no bacteria, (5) no adverse effects on the formation, such as clay swelling, (6) no detrimental effects when mixed with formation fluids, and (7) non-conductivity to prevent galvanic corrosion.

Because this fluid does not exist in nature, the waterflood engineer must strive to provide a water that is an acceptable compromise. The water source and treatment should be designed to produce a water that approaches this ideal as closely as economically feasible. The economic factor is critical because the best possible water may be too expensive, thus reducing waterflood profits.

The sources of water may be surface or subsurface and may include water produced with oil and gas. The important considerations are the rate and total volume required and the quality of the water. Those factors will determine the costs of achieving an acceptable water source.

Water-treatment systems affect the economic or physical success of a waterflood. There are many variables in design and equipment selection for water treatment. The system selected should consider the amount and quality of the water to be treated, the reservoir characteristics, and the waterflood economics. In some cases, when the life of the project is sufficiently short, corrosion control is not required, but further tertiary recovery plans must also be considered. In other cases, the reservoir transmissibility is so high that a relatively poor water can be successfully injected. All these conditions must be taken into account when the engineer designs a water-source and -treatment system.

4.2 Water Supply Source

The waterflood engineer must determine an acceptable water source early in the planning process. Sources may be broadly classified as surface or underground and fresh or saline. Almost any water can be made usable for waterflooding, although not necessarily with acceptable economics. The first source examined is often fresh water, but its use frequently is limited to humans, animals, or agriculture. In recent years, most waterfloods have used saline or other nonpotable waters, including ocean water. Partly because of environmental regulations, produced water is usually injected.

When underground water is to be considered, electric logs and other available data can be examined to find a potential supply. Any aquifers that provide an active waterdrive for other oil reservoirs deserve careful evaluation. Water production from such aquifers could cause an oil-production decrease as a result of a weakened drive mechanism. Fortunately, non-oil-productive water sources are often present near oil reservoirs.

4.2.1 Rate Requirements. Available rates of water must be adequate for daily injection needs throughout the operation's life. Design rates are usually specified by the reservoir study, but some generalities apply. For most of the flood life, an oil reservoir will require about 0.5 to 1 B/D [0.08 to 0.16 m^3/d] of injection water for each 1 acre-ft [1233 m^3] of reservoir volume. The ultimate volume of water required for many floods will be on the order of 1½ to 2 times the PV of the reservoir being flooded. Some of this water may be recycled produced water.

Data on source-water producing capacity must be obtained to evaluate a potential water source. For underground supplies, it may be advisable to complete and to test a water well, including any required stimulation treatment. For surface water sources, such factors as the historical water availability, water rights, other current or potential users, and seasonal changes need to be considered. If water is to be purchased from an outside source, these same factors should be evaluated for that system.

4.2.2 Produced Water as Source Water. Because many operators use produced water as part of their total requirements, this source is discussed first. Commonly, waterflood operations start with an external source and gradually change over to produced water as breakthrough occurs. It is normal to shut in producing wells as they approach water cuts of 100%. This reduces the amount of water needed to balance production. For this reason and because of economics, produced water often becomes the only source used late in the project life.

Produced water is not necessarily a good source. Reservoirs often produce water that is incompatible with other produced or source waters being used, which may cause such problems as undesirable precipitates or scale. Produced water also contains some oil, which may cause injectivity problems, as discussed later. Further, as the water is processed in the production and treatment systems, it is physically changed and frequently chemically altered. The physical variations may include changes in temperature and pressure, formation of emulsions, and inclusion or removal of such solids as sand or metal particles.

Chemical changes may include (1) reduction of pressure-releasing gases from solution, (2) introduction of corrosion products into the water, (3) introduction of oxygen into the water and subsequent reactions, (4) formation of scale or other precipitates, and (5) changes resulting from chemical treating by the operator, such as scale inhibitors, emulsion breakers, bactericides, or even well-stimulation treatments (acid or fracturing fluids).

These changes in the nature of the produced water may result in the need for additional treatment. Incompatible waters may be processed and injected separately, but this results in problems caused by the constantly changing ratio between the produced and source water. Proper handling of the produced water may reduce the need for subsequent treatment, as discussed below. Produced water needs to be handled as little as possible to minimize problems.

4.2.2.1 Elimination of Air Contact With Produced Water. When processing produced water, it is important to prevent exposure to air. Preventing contact with air is usually economically feasible.[1]

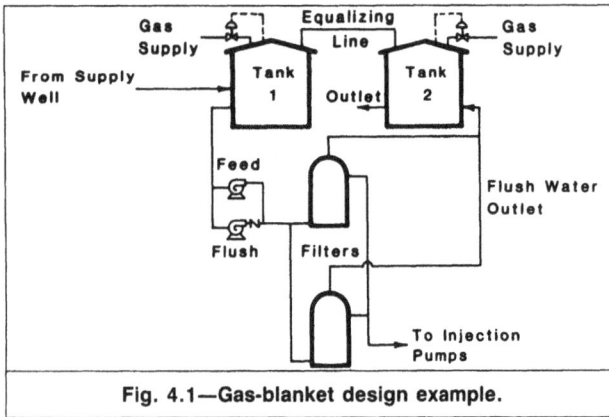

Fig. 4.1—Gas-blanket design example.

Aeration of produced waters has a number of results:

1. Oxygen reacts with dissolved chemicals in the water and produces oxide products, either as precipitates or in solution.

2. Oxygen increases the corrosivity of the water.

3. Deposition of precipitates and corrosion products provides a breeding place for anaerobic bacteria.

4. Scale and other precipitates can cause plugging in injection wells.

5. Air can carry sulfate-reducing bacteria into the water.

6. Changes in the dissolved gases can destabilize the water.

Air will be excluded if the entire production/injection system is kept above atmospheric pressure at all times. Care must be exercised to prevent pressure drawdown at pump suctions. Good design practices and the use of "gas blankets" will ensure that the system is kept air-free. Designing lines and wellheads to eliminate open casing annuli and low pressures caused by line elevation changes is desirable. Pump seals, packing, and the injection system can be designed and maintained to prevent air being sucked into the system. Gas blankets are systems that inject natural gas, nitrogen, or other inert gases into the system to maintain a pressure slightly greater than atmospheric. Each installation will have specific needs, but supply wells and tanks generally require gas blankets. Pressure vessels must either be kept fluid-packed or have gas blankets. The design of gas blankets is fairly simple, but some factors should be considered by the waterflood engineer. All regulators and lines can be sized to supply gas at rates high enough to maintain positive pressure when the fluid level in a tank drops at its maximum rate. Regulators should be installed as close to the gas entry point as practical. The system design preferably should eliminate or remove condensed liquids that could freeze and cause blockage of gas lines for makeup, control, or relief. An interruption in the gas supply, followed by a rapid decrease in the water level, can create a vacuum in the tank vapor space. A properly sized check valve should be provided to open the tank to the atmosphere, thus preventing damage from tank implosion.

Minimizing changes in tank levels is good design practice. One method of accomplishing this is to connect the vapor spaces of different tanks together (Fig. 4.1). Changes in the levels, and therefore the vapor spaces, in different tanks will frequently tend to cancel each other. Interconnecting several tanks also allows multiple en-

try points for gas into the blanket system. Connecting piping needs to be large enough to allow adequate movement of gas from tank to tank.

Every gas-blanket system will be different, but the principles outlined above will assist the engineer in the design process. The essential point is to eliminate air entry.

4.2.2.2 Oil Removal. Water and oil interfere with their respective movement in a reservoir. As a result, the produced water should be stripped of all oil that can be economically removed. The reservoir interference of oil and water is the result of relative permeability effects. In simple terms, this means that the maximum permeability to water (and therefore maximum injection rate at a given bottomhole pressure) exists when there is only water in the reservoir. Although residual oil remains in the reservoir, any oil added from injection will decrease the permeability of the zone to water.

This reduction will result in increased cost because of the additional power required to inject any given amount of water. Oil will coat any suspended solids that it contacts in the water. This will tend to reduce the effectiveness of any solids-removal treatment system and will increase costs. Oil that is injected with the water is generally lost and will not be sold. Various systems for oil removal are discussed later, but the importance of doing so is clear.

4.2.2.3 Dissolved Gases. The two gases dissolved in produced water that most commonly cause problems are H_2S and CO_2. H_2S or CO_2 in water generally increases the corrosion potential, frequently by a significant amount. In both cases, weak acids are formed and the corrosion that results causes pits to form. This type of attack can cause early and frequent failures in exposed steel lines and vessels.

H_2S may also result in hydrogen embrittlement of susceptible metals, causing premature failures. H_2S is a highly toxic gas that requires special handling procedures to ensure safe operations. The engineer should consult metallurgists and safety experts for design advice if significant quantities of H_2S are expected or possible in the waterflood project. Quantities of H_2S in excess of 20 ppm should be considered significant. This gas is lethal at low concentrations, depending on the length of exposure time.

Another problem associated with both H_2S and CO_2 is the formation of precipitates, such as iron sulfide and calcium carbonate, respectively. These precipitates can cause scaling in equipment, lines, and wells, as well as plugging in the wells and formation.

4.2.3 Surface Water. As shown in Fig. 4.2, fresh surface water is available from ponds, lakes, streams, and rivers. These sources are subject to possibly extreme seasonal changes in quality and quantity and contain varying quantities of organic and inorganic impurities. Generally, the water will be saturated with oxygen and CO_2 from the air. These dissolved-gas levels will vary with water temperature and recent rain activity. Another surface source shown in Fig. 4.2 is the ocean, especially for offshore or inland marine operations.

Although a surface supply may appear attractive at first glance, it is usually considered the "last resort" as a water source. On the other hand, water taken from underground adjacent to the surface supply can be an excellent choice, as discussed in Sec. 4.2.4. Water withdrawal directly from the surface supply is usually taken from the best level for good water quality. This level may range from

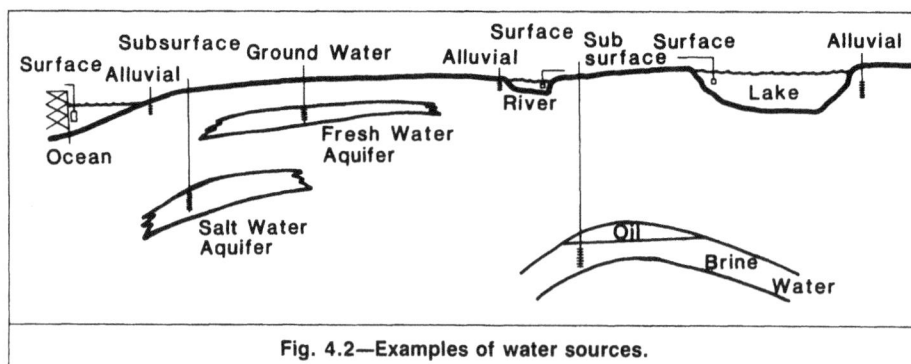

Fig. 4.2—Examples of water sources.

near the surface to near the bottom, depending on the conditions.

Water rights must be obtained to use fresh surface water. In the U.S. these rights are vested in various entities. In virtually all the eastern states, the landowner holds water ownership. In the western U.S. most water is held by the states, Indian tribes, or the federal government. Regardless of ownership, it is necessary to purchase or otherwise to obtain the right to use surface water in an injection project. This can be a significant cost for a waterflood. Use of ocean waters requires permits from federal and, sometimes, state agencies. This permitting process may be difficult, expensive, and time-consuming. Use of fresh surface water may also be restricted or shut off in years with low water flows. This last possibility can have serious consequences and must be kept in mind when planning to use surface water.

4.2.4 Subsurface Water Sources. *4.2.4.1 Potable Water.* In many places, alluvium beds are associated with streams and rivers (Fig. 4.2). These sand and gravel beds can provide an excellent fresh-water supply at low cost through shallow wells. Because the alluvium beds themselves act as filters, only minimal water treatment may be needed. The wells usually are shallow (< 300 ft [91 m]) and may be closely spaced because the alluvium bed is constantly charged by the river or stream. Even in dry season, when the surface river flow is very low or nonexistent, the subsurface system may still be charged with water.

Rock formations down to 1,000 ft [300 m] or even deeper may contain a supply of fresh water known as groundwater. Development of this water source is frequently costly. Extensive testing may be needed to determine deliverability and reserves. A draw-down test can be run by pumping the well continuously for several days to determine how much the producing water level is lowered. Rock formations vary in porosity and permeability. Some zones, with low values for these parameters, may produce all available water in a short time upon continuous pumping. Naturally, these wells are not suitable as a water source. On the other hand, if continuous pumping does not significantly lower the water level, a high-rate water well is indicated. The pumping rate is determined with a meter as the well is tested.

When water is to be produced from more than one well in the same formation, it is essential that the wells be properly spaced to prevent interference between them. If a large amount of draw-down is exhibited at the first well, chances are good that wide spacing may be required. Sometimes 2,000 ft [600 m] or more is needed between wells. Geologic data, reservoir behavior, well test information, and good judgment are required to space wells correctly. Because potable water sources are generally used by others, information on well tests, production histories, and well spacing will often be available from state and federal regulatory agencies.

4.2.4.2 Nonpotable Sources. As shown in Fig. 4.2, saline water is usually present in strata lying closely above or below the oil-producing formation. It is sometimes possible to use nearby dry holes or abandoned wells to produce from a water-bearing formation. This possibility should be checked if there are wells located favorably for the project. Aquifers should be checked to ensure that production from them will not affect oil production from adjacent fields. Electric or other logs from producing wells can help to locate saltwater sands and to trace sand continuity from well to well. Thick sands with continuity over a large area will usually be the best water sources. The test procedure described for freshwater wells can also be used to determine the deliverability of saltwater wells. Because salt water is normally found deeper than fresh water, the lift is greater and the costs are usually much higher. Source wells may require stimulation by hydraulic fracturing or acidizing to increase productivity or to maintain the production rate.

Shallow wells near oceans may tap an unlimited saltwater supply. High-capacity production often comes from caissons or galleries instead of conventional wells. This water may need less treatment than water taken directly from the ocean. Water may also be drawn from beneath the ocean floor, allowing the bottom sediments to act as a filter. The large Wilmington field[2] uses water from the Gaspur zone as a supplemental source to reinjected produced water. This formation outcrops on the ocean floor and supplies essentially filtered ocean water. However, extensive water treatment may be required for any subsurface salt water.

4.3 Water Quality

Required water quality varies from flood to flood. The "best" water is the least costly that can be injected at satisfactory rates and does not cause mechanical or chemical problems. In this monograph, "water quality" refers to those properties of the water that can cause effects on the reservoir, wells, and surface facilities that may lead to the economic success or failure of the waterflood; water quality is not used here in the sense of potability or its effect on the environment. This section discusses monitoring of such properties as chemical reactivities, corrosion potential, and plugging tendencies.

The cost of good water quality will include such factors as water treatment, corrosion control, operating costs, and incremental development costs. If more than one source is available, a cost-comparison study is appropriate to decide which water is best. For example, extensive studies for the Kuparuk and Prudhoe Bay waterfloods determined that water from the ocean was less costly than the available subsurface waters. Laboratory and/or field injectivity tests are needed to evaluate water-quality requirements properly.

4.3.1 Sampling. Source-water analyses will usually reveal potential quality problems. Water samples must be collected under the producing conditions of the source water. The sample containers must be clean and the water not exposed to air. Laboratory water-analysis personnel or other experts should be consulted to determine the best sampling techniques. All samples should be taken at least in duplicate. Contaminated samples can lead to erroneous conclusions about the treatment needed. Samples must be fresh or properly preserved when delivered for chemical analysis.

A standard laboratory water-analysis report[3] is shown in Table 4.1. At a minimum, a water analysis includes[4] the concentration of such dissolved ions as sodium, calcium, magnesium, barium, iron, chlorides, bicarbonates, carbonates, and sulfates. Specific gravity, resistivity, and pH should also be measured. Suspended solids, oil and grease concentrations, and dissolved gases may also be reported but are generally best tested on site. These analyses will give important information on the scaling tendencies and corrosivity of the water. The waterflood engineer can then design the project to mitigate such effects in an economical manner. The exact design plan will depend largely on the specific reservoir and surface facility conditions.

4.3.2 Compatibility. If a sample of oil reservoir formation water is mixed with a proposed source-water sample, a precipitate may form, and the two waters are said to be incompatible. Two waters are said to be compatible if no precipitate is formed when they are mixed. Ostroff[5] discusses compatibility of waters and a method of testing. If source and formation waters are incompatible, it must then be determined whether source water is suitable for flooding. Incompatibility can create severe plugging and scale-deposition problems when the waters are mixed. If they are first mixed in the reservoir, the waterflood will not be affected until breakthrough occurs. At that time, the producing wells and equipment may begin to scale up and operating can become difficult and costly. As the flood progresses beyond this point, the precipitation and scaling problems constantly change because of the varying ratio of the waters being handled. Treatment should be considered to solve the problems. A common alternative is to have separate handling and injection systems. A comparative cost analysis is used to decide whether the source water should be used or how it should be handled.

Some waterfloods are very successful, even though the source water is not compatible with formation water. Bilhartz,[6] one of the first to point this out, based his conclusion on practical field observations of successful waterfloods where the water analyses showed incompatibility. A waterflood is frequently successful because the waters are first mixed in the pore spaces of the reservoir. If the source water reaches a producing well in one zone while another zone in the same well makes formation water, scale problems will likely develop rapidly. Similarly, breakthrough in some wells can lead to mixing the two waters in the surface equipment and consequent problems in production processing and, possibly, the injection system and wells. Typically, a waterflood that uses incompatible waters will experience no difficulties until breakthrough; then, se-

TABLE 4.1—API WATER ANALYSIS REPORT FORM

Company		Sample No.		Date Sampled
Field	Legal Description	County or Parish		State
Lease or Unit	Well	Depth	Formation	Water, B/D
Type of Water (Produced, Supply, etc.)		Sampling Point		Sampled by

Dissolved Solids
Cations   mg/L   me/L

Na (calc.)   _____   _____
Ca   _____   _____
Mg   _____   _____
Ba   _____   _____
_____   _____   _____
_____   _____   _____

Anions
Cl   _____   _____
SO_4 _____   _____
CO_3 _____   _____
HCO_3 _____   _____
_____   _____   _____
_____   _____   _____
_____   _____   _____

Total dissolved solids (calc.)
                   _____

Fe (total)   _____
Sulfide as H_2S _____
Remarks and Recommendations

Other Properties
pH   _____
Specific gravity, 60/60°F   _____
Resistivity (Ω-m)_____°F   _____
_____   _____
_____   _____
_____   _____

Water patterns (me/L)

In this table, calc. = calculated and me = milliequivalents.

vere scale problems that gradually decrease as the waters reach equilibrium will occur. This process may or may not preclude a successful waterflood, depending on severity.

If possible, compatibility tests are to be conducted on site with fresh or in-line samples of the waters involved. Testing should simulate expected waterflood conditions as much as possible. Mixing of incompatible waters can lead to serious consequences in both surface and subsurface systems, so extensive testing (if needed) is justified.

4.3.3 Formation Sensitivity. Some reservoir rocks, particularly sandstones, suffer permeability reduction when contacted by injected water. Damage is usually associated with relatively low-salinity injection water. This formation sensitivity and the resultant permeability damage has two main causes that may both be present.

1. Water reacts with the indigenous clays, causing them to swell or to disperse and thus to constrict the pore throats.

2. Water flows through the rock, dispersing and rearranging fine formation particles to cause blocking or plugging of the pore system. Reductions in salinity can cause clays to lose their cohesive forces and to start to move through the pores.

Several tests are available to help determine the presence and degree of formation sensitivity. Generally, these tests require suitable core samples of the reservoir rock. If such cores are not available and if formation sensitivity is suspected, it may be economical to drill a well to obtain core samples. An alternative is to inject water of decreasing salinity to test the effect on injectivity. Hewitt[7] recommends that four tests be run on core samples: (1) core-injection tests to measure permeability vs. water salinity, (2) X-ray diffraction analyses to detect clay swelling, (3) tests to determine the swelling capacity of indigenous clays, and (4) microscopic examination of rock thin-sections to determine clay mineralogy and distribution. Also useful is the core-injection test, conducted to detect nonswelling clays that disperse and cause permeability reductions.

When evidence indicates that the reservoir rock is subject to damage by contact with the injection water, several alternatives are available to mitigate or to eliminate the formation damage. The obvious solution is to change to an injection water that will not cause swelling or dispersion; however, this is frequently not possible. A number of investigators have reported[8-13] various techniques to stabilize indigenous clays. Most of these solutions involve chemical treatments that react with the clays or coat the clay particles to prevent exposure to the water. The state of the art in clay stabilization is constantly evolving. Most reservoirs are not water sensitive and will not suffer from permeability damage even if fresh water is used.

Because few new waterfloods use fresh source water, the subject will not be discussed further in this monograph. If the waterflood engineer suspects formation sensitivity, it is recommended that a current literature search be made on clay-stabilization techniques. Available experts in this area should be consulted and the best current technology used to mitigate the problem.

4.3.4 Water-Quality Monitoring. Water quality can be monitored by discrete, individual, or continuous sampling methods. The data collected usually include chemical water analyses (Sec. 4.3.1), suspended solids size distribution, total solids quantity, type of solids, oil and grease concentration, corrosivity, and oxygen content. Sampling points should be selected throughout the water system to allow observation of any changes in the quality. Suggested points are at the water source, upstream and downstream of treatment and injection pump facilities, and at the injection wells.

Discrete sampling methods include periodic membrane filter tests and water sample analyses (see Sec. 4.3.1). The former test consists of flowing the water through a membrane filter, measuring rate and time, and analyzing the material trapped on the filter pad. This technique will yield such information as suspended solids and oil concentrations, nature of the suspended solids, and the plugging rate. Tests should be performed regularly to detect any changes in water quality. Complete instructions for using membrane filters are found in the literature.[14-16] Barkman and Davidson[17] discuss the theoretical analysis of membrane filter data. Cerini[18] presented empirical correlations between filter data and water quality, and McCune[19] has discussed on-site testing for water quality. The methods described in these papers are valuable guidelines to consider, but actual reservoir behavior may differ considerably. These tests are most useful to show trends in water quality or the differences between waters. Other factors that affect reservoir plugging are particle size and distribution compared with pore-throat sizes. A rough rule is that plugging usually will not occur if the particle diameter is less than one-third of the pore-throat size as defined by the effective hydraulic diameter. Instruments such as Coulter counters can give particle-size-distribution data on discrete samples. Solid concentrations can be measured in some waters by turbidity-measuring devices on a continuous basis. These instruments operate on the principle of light-scattering by the particles in the water and hence depend on the water transmitting the light effectively. Systems that have considerable oil or other materials entrained in the water can severely affect the operation of these devices. A new instrument that uses the reflection of a focused ultrasonic beam off solid particles in the water has been developed to measure solid concentrations on a continuous basis without being affected by opacity,[20] although oil droplets do change the measured values.

The most common method of monitoring corrosion rates in waterflood installations is the use of coupons. These are pieces of metal, usually bare steel, inserted into the water stream. At periodic intervals, the coupons are replaced and the old ones weighed. The weight loss is then converted to a corrosion rate. The Natl. Assoc. of Corrosion Engineers has published a recommended practice[21] on the use of coupons. They can be of various sizes, shapes, and weights. For example, an entire tubing joint can be cleaned, weighed, and run into an injection well. The next time the well is pulled, that joint is removed, examined, and weighed to determine the type and rate of corrosion. In this case, the entire tubing joint is a corrosion coupon. Because this is not always practical, a short piece of tubing called a "sub" is frequently used in place of a full joint. Watkins[22] discusses the use of cylindrical corrosion coupons inserted through small connections into the water stream. More frequently, the coupons are flat plates of various shapes and sizes.

A method of continuously or periodically measuring corrosion using electrochemical principles has been developed. The method uses probes to measure the electric resistance in a wire exposed to the water compared with an identical wire at the same temperature but not exposed. As the wire corrodes, the resistance changes. These probes now give accurate readings from <1 mil/yr [25.4 μm/a] up to several hundred mils per year.[23] Sour systems can cause problems because iron sulfide may coat the probe. The instrument will then give erroneous measurements. However, these devices have several advantages. They give corrosion-rate readings in a much shorter time than coupons, and data can be collected over time without removing the probe. Thus, changes in rate can be plotted. This may have special advantages in inhibited systems, where the rate is high initially, decreases as the inhibitor film builds up, and finally increases again as the film deteriorates.[24] Using the probe in this way can help determine how much and what

type of corrosion inhibitor is required, as well as whether batch or continuous treatment is optimal. Another system has been developed that uses the electrochemical cell potential to measure corrosion rates. The engineer should check current information to ensure that the best practical instruments are being included in the design.

Every waterflood system will require some monitoring of water quality. The best procedure for an individual project will depend on the water, facility, and reservoir characteristics. Using the information on needs and equipment outlined above, the waterflood engineer can design the most economical system for water-quality monitoring for his project.

4.4 Types of Water Treatment

The equipment and processes available for treating water are constantly evolving and improving. The waterflood engineer must keep current through reading literature, attending seminars or conferences, and staying in contact with industry personnel working in this area. Without this effort, the engineer may find his or her knowledge obsolete in 1 or 2 years. Ostroff's[25] *Introduction to Oilfield Water Technology* is a classic in this field and should be obtained as a reference. Another good reference is Patton's[26] *Oilfield Water Systems*. Various vendors of water-treating chemicals or equipment can provide books and brochures on water-treating principles and systems.

This section lists common water-treating problems and suggests possible solutions. The specific types of equipment mentioned may be eventually replaced by better designs, but the waterflood engineer should ensure that the latest and most suitable technology available is used when designing a water-treatment system. It is emphasized that the best possible waterflood project requires *no* water treating. The engineer should make every effort in designing water production and injection systems to eliminate the need for water treatment. The engineer should use water treatment only if this effort fails as a result of circumstances or economics. The design principle is to minimize the treatment system.

4.4.1 Dissolved Oxygen Removal. Surface source waters will normally be saturated with oxygen. Other source waters may take oxygen into solution during processing. As part of a water-treating system, removal of this dissolved oxygen is usually desirable. Battle[27] gives the following methods of oxygen removal: (1) chemical reactions with sodium sulfite, hydrazine, or sulfur dioxide; (2) vacuum deaerator; and (3) countercurrent stripping with natural or inert gas.

Chemical reaction will remove small amounts of oxygen by a reaction with an easily oxidizable chemical, such as sodium sulfite, sodium and ammonium bisulfate, or hydrazine,[25] called an oxygen scavenger. In seawater systems, the water frequently is chlorinated upstream of the deaerators. Any residual chlorine will react with the scavenger before the oxygen, increasing the use and cost of the scavenger. The engineer should design for a minimum residual while allowing an adequate amount for chlorination. Scavengers are commonly used after vacuum deaeration or countercurrent stripping to reduce the oxygen content from a few parts per million (1 to 3 ppm) to <0.05 ppm. Sodium sulfite is cheaper than hydrazine and is more often used in waterflood applications. A catalyzed form is normally needed to accelerate the reaction, especially in water containing CO_2. Some waters contain natural catalysts and require no additions. If needed, a catalyst, such as cobaltous hexahydrate, may be purchased separately and added in the field to decrease the cost. Liquefied sulfur dioxide, purchased in bulk quantities, may be competitive for large-volume systems. Some operators, however, find that generation of sulfur dioxide by the burning of sulfur is less costly than purchase of the chemical.[28]

Oxygen removal by vacuum deaeration has not always been successful in the field. The failures were probably caused by the relatively inefficient and expensive vacuum pumps of earlier days. Adams[29] states that with the efficient and economical vacuum equipment now available, this should no longer be a drawback. A typical vacuum deaeration unit is shown in Fig. 4.3. A single stage is shown, although multiple-stage designs yield a lower oxygen con-

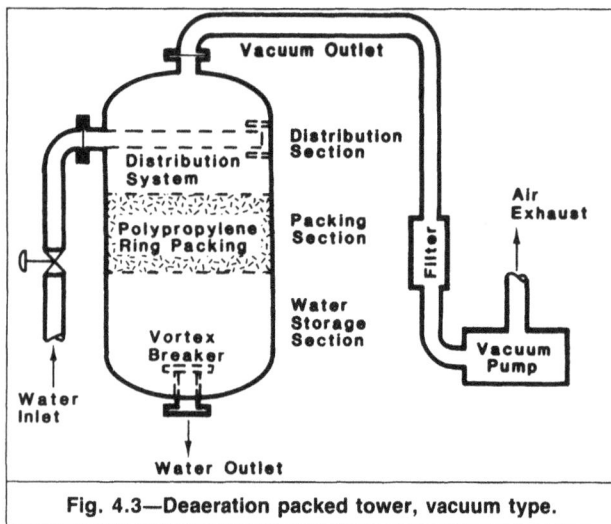

Fig. 4.3—Deaeration packed tower, vacuum type.

tent. For this reason, three-stage units have been used in several waterflood operations.[30] Because the solubility of a gas in a liquid is directly proportional to the pressure of the system, reducing the pressure reduces the amount of dissolved gas. The packing in the deaerator and the correct flow rate allow a larger surface area for the water, permitting it to reach equilibrium conditions as it passes through the tower. For design purposes, the water flow rate is on the order of 40 to 50 gal/(min-ft^2) [0.027 to 0.034 m^3/(s·m^2)] of cross-sectional area.[25] The packing must not become flooded, and channeling of the packing should be minimized by proper packing distribution. The vapor-removal lines must be large enough to prevent excessive pressure drop. A vacuum deaerator should be seriously considered for oxygen removal where hydrocarbon or inert gases are unavailable or too costly.

Countercurrent stripping of oxygen from water by natural or inert gas is another method of oxygen removal.[31] If a good-quality natural gas is available—i.e., little CO_2 and H_2S—this process may be attractive, because the effluent gas from the deaeration process can be used in fuel-gas systems. Figs. 4.4 and 4.5 show the simplicity of the process. The operation depends on balancing the flow rate of gas moving upward in the column against the water rate moving downward. The water rate must not be so great that it decreases the oxygen-removal efficiency. In one example,[32] about 1.75 ft^3 gas/bbl water [0.31 m^3 gas/m^3 water] is needed to reduce the oxygen content from 10 to about 0.1 ppm. At the Kuparuk field in Alaska, the deaerators (Fig. 4.4) were designed to reduce the oxygen content from 15 ppm to 20 ppb using 4 ft^3/bbl [0.7 m^3/m^3]. Weeter[33] gives the details of designing a countercurrent

oxygen-stripping system. A combination design that uses countercurrent stripping or vacuum deaeration, or both, has been developed. This system was chosen for the Prudhoe Bay seawater treating plant.

If CO_2 is present in the water, it will also be released as the oxygen is removed in a vacuum system. Conversely, if the natural gas is rich in CO_2, the water will absorb the CO_2 in a stripping tower. Either process may be good or bad, depending on the type of water being treated. Removal of CO_2 causes an increase in the pH of the water, which may result in the formation of calcium carbonate scale. Absorption of CO_2 will decrease the pH and reduce the possibility of scale. Corrosion, however, may result from too much absorption of CO_2.[34] The engineer should be aware of these effects.

4.4.2 Acid Gas Removal. The acid gases, H_2S and CO_2, usually result from oilfield production, although CO_2 can be absorbed into the water from the air at specific conditions. There is no simple chemical treatment that will remove these gases, although acrolein has been used successfully in a number of cases.[35] Therefore, some mechanical means is normally preferred. Handling these gases after removal may present problems. For example, H_2S releases may not be allowed because of clean-air regulations. Removal of these gases is usually expensive. The more common solution is to incorporate design features that minimize the effect of the dissolved gases. This approach is generally more economical than removing the gases, particularly if the removal process is aeration. The waterflood engineer should examine the comparative economics to determine whether designing for a sour system or reducing the gas concentrations is preferable.

One system to remove acid gases is to aerate the water, which then leads to the problem of subsequently removing the oxygen. Aeration also causes the iron and manganese to be oxidized and precipitated. Ostroff[37] describes nine different types of aerators. Three commonly used in waterflooding systems are the coke-tray, the wood-slat-tray, and the forced-draft aerators. In each of these, the basic principle is to ensure thorough contact of the water with air.

Weeter[36] has described the design of a packed column tower for the removal of H_2S by use of cooled exhaust gas. Sweet gas may be used in this type of system instead of flue gas.[37] Oxygen is not purposely introduced with this process, although some will be contained in the flue gas. For installations in which large volumes of water are handled, this cooled exhaust gas or sweet gas process is well-suited for H_2S removal.

The lowest CO_2 or H_2S concentration that can be reached by aeration depends on the relative solubilities of these gases to air and on the characteristics of the water. Significant reductions of the unwanted gases usually can be obtained down to a few parts per million.

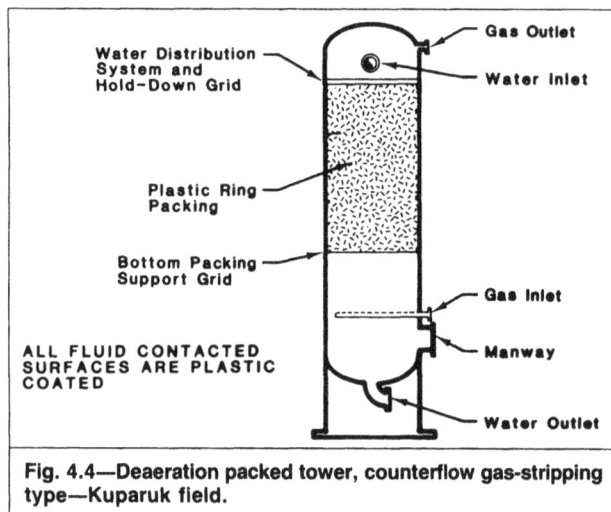

Fig. 4.4—Deaeration packed tower, counterflow gas-stripping type—Kuparuk field.

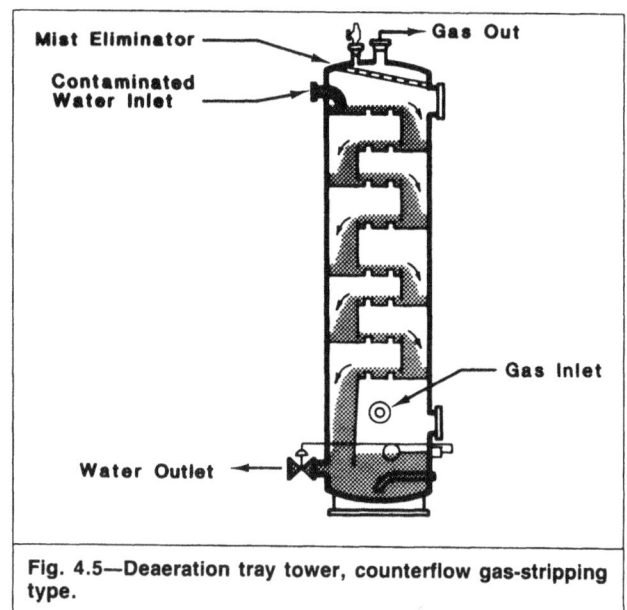

Fig. 4.5—Deaeration tray tower, counterflow gas-stripping type.

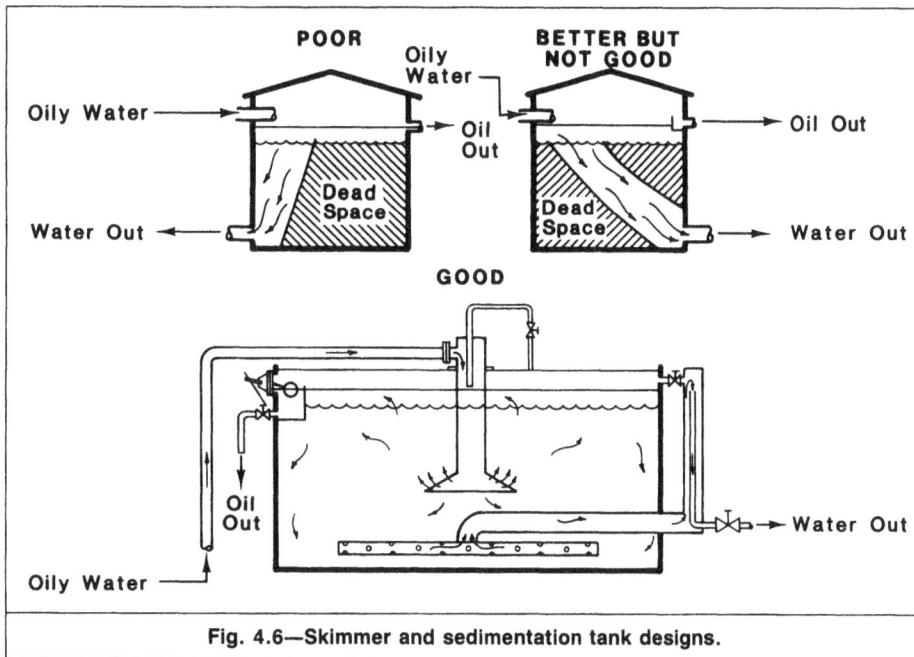

Fig. 4.6—Skimmer and sedimentation tank designs.

4.4.3 Suspended Solids and Oil Removal. Solids in the water can cause problems—such as plugging of lines, valves, or other surface facilities; erosion in valves, pumps, and lines; and wellbore and formation plugging. Solids may also precipitate as scale. Scale control will be discussed in Sec. 4.4.4. Solids that tend to plug the formation are primarily particles of sand, silt, clay, and organic matter. Precipitates formed by chemical reactions and the products of corrosion, such as iron oxide or iron sulfide, are also possible plugging agents. Biological agents, such as bacterial colonies, algae, and plankton, may also present a problem, particularly in surface source waters.[38]

Removal of suspended solids is usually required when they exceed 1 to 5 ppm, depending on the formation. These particles can be removed mechanically. Probably the simplest method of removing solids from water is the process of sedimentation. Sedimentation is accomplished by providing a pond or tank of sufficient size and correct design so that solids in water slowly flowing through the pond will drop from the water and settle to the bottom.

Sedimentation ponds, tanks, or basins may be constructed as walled ponds, earthen pits, or wood or metal tanks. These are usually designed with a sloping bottom to facilitate sludge removal. Dual systems are frequently installed so that sedimentation can continue during cleanout periods and repairs. The ponds or tanks often include baffles to eliminate dead spots in the pond and to increase the retention time to allow more effective settling. Retention time is the time required for a unit volume of water to flow through a sedimentation basin or tank. The amount of retention time required is governed by the rate at which the solid particles will fall through the water to the bottom of the basin or tank. This rate can be theoretically calculated by Stokes' law:

$$v = \frac{gd^2(\rho_w - \rho_o)}{\mu_w},$$

where

  v = rate of fall,
  g = acceleration of gravity,
  d = diameter of particle,
  ρ_w = density of water,
  ρ_o = density of particle, and
  μ_w = viscosity of water,

with all quantities in consistent units. Some of these factors are not usually known accurately, but the basic relationships are valid and approximations can be made. As a rule of thumb, a true retention time of 2 to 4 hours and an average flow velocity of 1 ft/min [0.5 cm/s] will provide adequate sedimentation. Theoretical retention time may be calculated by dividing the effective basin or tank volume by the throughput rate. Because of the effects of short-cutting flows and dead spaces, however, this calculation will virtually always indicate a settling time much longer than actual. Zemel and Bowman[39] discuss techniques for determining true retention times in similar systems. In many cases, the addition of chemicals may increase the effectiveness of the sedimentation process. Coagulants, such as aluminum sulfate, potash alum, ferric/ferrous sulfate, ammonia alum, and various polymers, can cause smaller particles to flocculate. This larger effective diameter then causes the particles to fall faster, per Stokes' law. The choice of chemicals to be used in this manner should be determined by on-site testing because many factors affect this reaction.

Small amounts of pure oil in an injection system rarely cause plugging problems unless a large amount of asphaltenes, paraffin, or other hydrocarbon solids is present. However, suspended oil is usually found coating the solids in the water system. This tends to decrease the effective density of the particles, thus reducing the effectiveness of the sedimentation process. For this reason and because of the economic loss of injected oil, it is generally a good idea to remove the oil in the water-treating process. Sedimentation basins and tanks with skimming equipment can effectively remove free oil from the water. Fig. 4.6 shows several design examples for skimming/sedimentation tanks demonstrating problems and partial solutions for short-cutting and dead space. The process also behaves according to Stokes' law, but the oil droplets float to the surface. A number of commercial flotation devices are available, including corrugated plate interceptors, API-type separators, dissolved-gas flotation cells, and induced-gas flotation cells. Studies[40-47] have been published on the principles and effectiveness of this type of equipment. Fig. 4.7 shows a schematic of a gas flotation cell.[48] While the exact process involved in induced-gas flotation is not precisely known, the gas bubbles apparently attach themselves preferentially to oil droplets and solid particles. According to Stokes' law, this increased diameter and decreased density cause the droplet or particle to rise rapidly to the surface, where it is skimmed off by mechanical means. The current state of the art appears to be a settling/skimming vessel followed by an induced-gas flotation cell. This combination uses the principle of Stokes' law to remove oil and solids by both sedimentation and flotation processes. On-site pilot testing of the water to be treated is strongly recommended to ensure that the equipment chosen and installed will perform economically and effectively. Pilot testing also will provide an opportunity to determine the most effective types and amounts of chemicals to be used in the system.

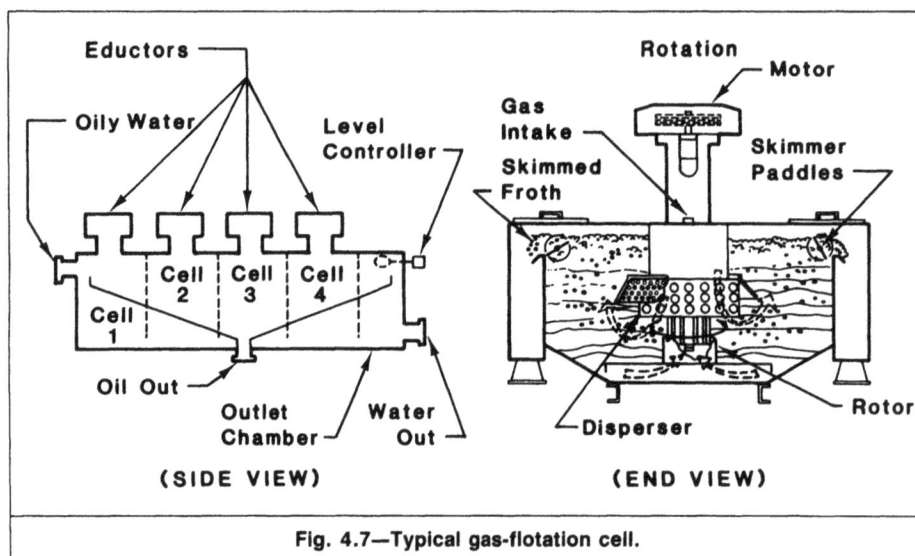

Fig. 4.7—Typical gas-flotation cell.

4.4.3.1 Suspended Solids Removal by Filtration. Many varieties of filtration equipment are available for the removal of suspended solids from water. The two basic types are gravity and pressure filters. As the name implies, gravity filters allow water to flow downward under the force of gravity. The filtration rate is slow, and the filters tend to be very large. This type of filter is rarely used in waterflood operations.

The pressure filter is commonly used in waterfloods. One general type uses surface filtration and includes cartridge and diatomaceous-earth (DE) filters. As seen in Fig. 4.8, cartridge filters are vessels that contain elements that can be regenerated or replaced. These elements are graded on the basis of the particle size that will pass through, ranging from 1 to several hundred microns. Generally,

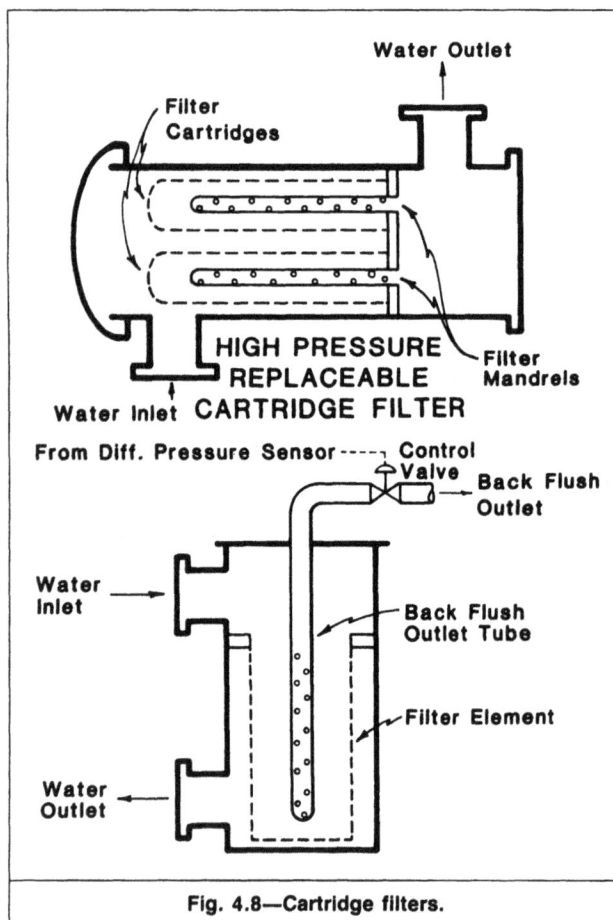

Fig. 4.8—Cartridge filters.

the elements are hollow cylinders made of various materials that fit over a perforated central mandrel. The flow is from the outside to the center. This type of filter is most commonly used as a polishing filter and is often installed at individual wellheads. For source waters with relatively low total suspended solids (TSS), such as seawater and subsurface waters, however, cartridge filters may provide effective and inexpensive primary filtration.[49,50] Such systems should be designed with at least one spare filter to allow a constant flow of water during backwash cycles or replacement of filter elements. If the TSS loads are continuous and exceed 1 to 2 mg/L, cartridge filters are not suitable because of short runs, high maintenance and operational costs, and frequent upsets. Disposal of used filter elements can also be a problem, especially in remote or offshore locations.

Another surface filter is the DE type, which uses a filter cake of DE applied to screens of various sizes, shapes, and materials. The screens or supports are contained within a pressure vessel through which the water flows. This type of filter can handle continuous loads of up to 20 mg/L TSS and provides excellent water quality. These filters can handle rates up to 5 gal/(min-ft^2) [0.003 m^3/(s·m^2)] at low TSS loads, but normally operate around 1 to 2 gal/(min-ft^2) [0.0007 to 0.001 m^3/(s·m^2)]. The filters are regenerated by removing the DE coating by backwashing and then applying a new coating. This procedure creates problems in supplying new DE and disposing of the old material, which will be more critical in remote or offshore areas. DE filters are also very susceptible to oil contamination. The most common application of DE filters is in preparing water for injection into low-permeability reservoirs where removal of very fine particles is needed.

The second general type of pressure filter uses bed filtration. These devices may have the water flow up or down through the bed. Allowable flow rates for these filters range up to 20 gal/(min-ft^2) [0.014 m^3/(s·m^2)] with little loss of efficiency. Actual rates will depend on the nature of the inlet water and the desired outlet quality. Some operators limit the allowable rate at little more than 50 to 60% of maximum. Single or multiple types of media are used. The vessels have water-distribution systems to spread the flow evenly over the surface. The filter may be horizontal or vertical. Because of the depth of the beds (3 to 6 ft [1 to 2 m]), these filters have much greater holding capacity than the surface filters. Their run cycles are much longer and can be up to several days. Figs. 4.9 and 4.10 show schematics of the upflow and downflow filters.

The upflow vertical filter (Fig. 4.9, after L'eau Claire Systems Inc.) uses a distribution plate with a large number of carefully designed nozzles to distribute the water flow evenly. The water flows upward through a layer of gravel and then through two different sizes of silica sand. The grid between the two grades of sand prevents excessive mixing. The top grid prevents significant overflow of the sand through the outlet. The sand builds compression arches under this grid during filtering operations. Other filter configurations do not always use these grids. The vacant head space allows

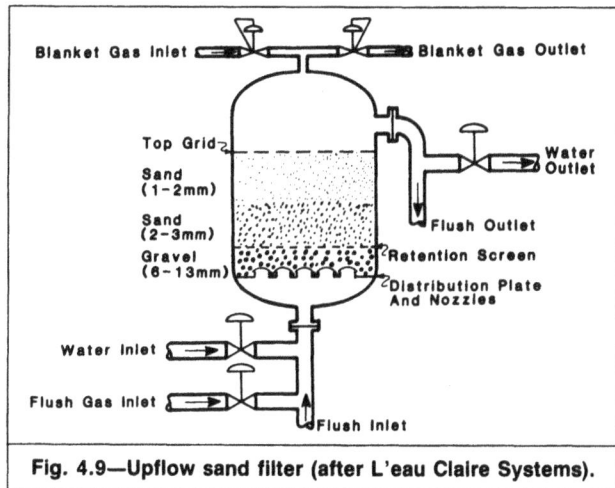

Fig. 4.9—Upflow sand filter (after L'eau Claire Systems).

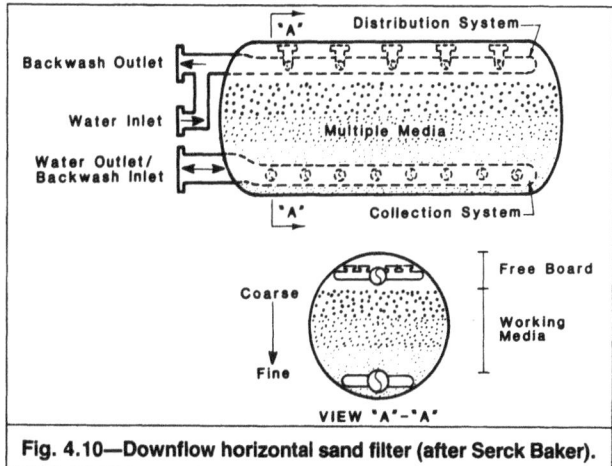

Fig. 4.10—Downflow horizontal sand filter (after Serck Baker).

expansion of the bed during backflushing. At least one design incorporates an oil-collection funnel to remove coalesced oil droplets as well as solids. Filter rates are usually around 8 gal/(min-ft^2) [0.005 m^3/(s·m^2)].

The downflow filter shown in Fig. 4.10 (after Serck Baker) is a horizontal, multimedia vessel. Filtration rates may approach 15 to 20 gal/(min-ft^2) [0.01 to 0.014 m^3/(s·m^2)], although lower rates are often preferable. Evers[51] discussed mixed-media filters that were described as being particularly effective on water with suspended oil. These vessels are vertical downflow filters similar in shape to the upflow filter in Fig. 4.9. The media are several materials, each of different size and density, arranged in a coarse-to-fine manner in the direction of flow, thus providing greater filtration efficiency, higher flow rates, and improved storage capacity for the material to be filtered. A typical mix of three materials is 50 to 55% anthracite coal of 1.55 specific gravity, 25 to 30% silica sand of 2.6 specific gravity, and 10 to 15% garnet or ilmenite of +4 specific gravity. The sizes vary from 0.04 in. [1 mm] for coal down to 0.006 in. [0.15 mm] for the high-density material.

Both upflow and downflow filters of the types shown have been very successful in a number of waterfloods. After extensive pilot testing, the Prudhoe Bay waterflood used two kinds of filters but on different water sources. The produced water was treated most effectively in upflow sand filters while downflow, multimedia filters were the most economical means to treat the seawater used for a source water. These types of filters generally require the use of chemicals to achieve the most effective filtration. Chemicals such as alum and various polyelectrolytes are normally used. The waterflood engineer should plan on-site pilot testing to determine the optimum type of filter chemical treatment. The manufacturers of this equipment have skid-mounted pilot systems available and are very cooperative in arranging such testing. The investment in equipment is significant in those waterfloods that require filters. On-site pilot testing using the water to be treated is essential to make the best choice. The various factors to be considered in evaluating filtration equipment are discussed in the literature.[52,53] Determination of the water quality, source-water quality, and location (e.g., remote, offshore, or arctic) requirements will usually allow elimination of several types of filters. The remaining two or three should be evaluated on an economic basis. Those that remain viable should be put into a pilot test. Even if the waterflood engineer is able to determine that only one type of equipment is best, it is strongly recommended that the chosen system be pilot tested in the field.

4.4.4 Scale Prevention. When insoluble compounds are formed in water from chemical reactions of constituents in the water, these may be deposited on pipe or other equipment contacted by the water. Depending on the components in the water, some of the precipitates that may form scales in waterflood operations are calcium carbonate, magnesium carbonate, calcium sulfate, barium sulfate, strontium sulfate, and iron compounds such as ferric oxide and ferrous sulfide. The buildup of these scales may reduce the ID of the water-injection lines, cause meters and other equipment to become inoperative, and even deposit materials in the injection well, caus-

ing plugging. Consequently, an important part of water treatment for waterflooding is scale control. This is a complex chemical problem requiring the assistance of a competent water-treating specialist to recommend the treatment necessary to prevent scale formation. It is not wise to wait until scale becomes a problem to decide on control. Several papers[54-56] describe methods for predicting scaling tendencies on the basis of water analyses. For waters that are to be mixed, laboratory testing of various ratios and conditions is recommended.

Where scale is a potential problem, sequestering agents or scale inhibitors are used to prevent its formation. The inhibitor stops the growth of the crystal before it becomes large enough to precipitate. Many chemicals are available for inhibition of the various forms of scale, and more are being developed. The choice of the correct agent, the quantity required, and the manner in which it is used requires the knowledge of an experienced specialist for each application.

4.4.5 Slime and Bacteria Control. Slimes and bacteria are controlled by chemicals that kill the bacteria or at least keep them from multiplying. Bacteria can plug water-injection wells, water-distribution lines, and equipment such as filters. In addition, they often contribute to corrosion. Virtually all water systems contain some bacteria. The amount, type, and activity of the bacteria present will determine whether they are a problem.

Both fresh water and brines may contain bacteria in quantities that cause severe problems. If the waterflood engineer suspects that bacteria may cause plugging or corrosion, water samples should be tested by a competent laboratory to determine the type of bacteria present. Four kinds of bacteria commonly are found in oil fields: sulfate-reducing bacteria (SRB), iron bacteria, slime-forming bacteria, and *Clostridium*.[25] The API has established a standard[57] for testing for bacteria in water samples. Recent studies[58] indicate that the amount of sessile-type bacteria in a system may be several orders of magnitude higher than the amount found in the water. In a waterflood in the Java Sea,[59] it was determined that a new type of coupon called the Robbins device[60] gives more representative measurements than the standard API SRB bottles.

In addition to the amount and type of bacteria, the activity of the bacterial population will help determine whether an operational problem will develop. Chen and Chen[61] discuss some new methods of examining the metabolic status of the bacterial colonies in a water system. New methods of monitoring the amounts, types, and activity level of a microbial population are being developed. The waterflood engineer should consult the literature and specialists in this technology when designing a monitoring system or procedure.

Many types of treatment are available to control bacteria in water systems. Unfortunately, no one method is universally effective. Any treatment must contact the bacteria to kill them. Many water systems have large amounts of bacteria sheltered under scale or corrosion products. For effective treatment, these systems must be cleaned thoroughly. A procedure that has been satisfactory in many cases is scraping and acidizing followed by solvent or detergent

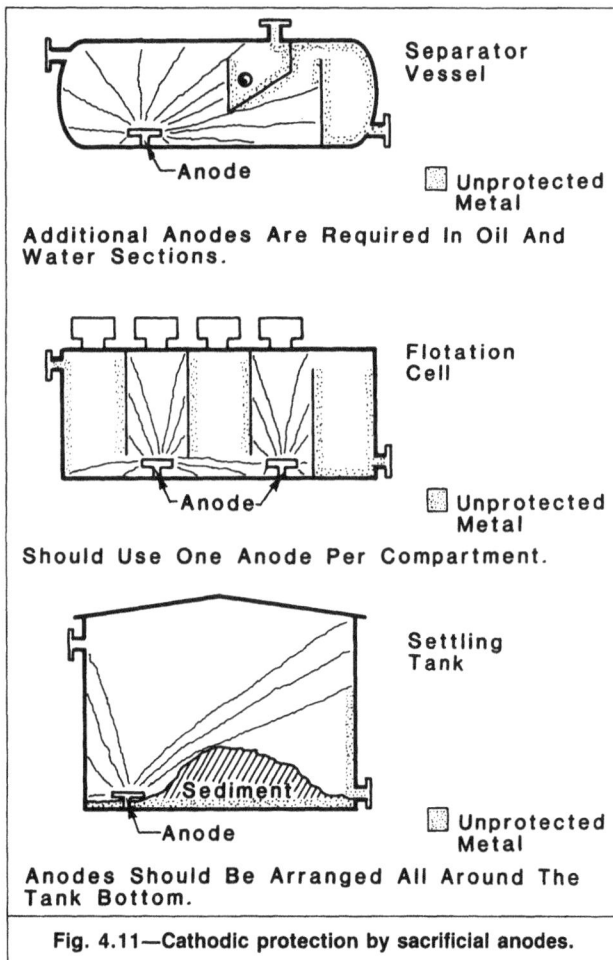

Additional Anodes Are Required In Oil And Water Sections.

Should Use One Anode Per Compartment.

Anodes Should Be Arranged All Around The Tank Bottom.

Fig. 4.11—Cathodic protection by sacrificial anodes.

soaks and flushes. The materials used will vary from one system to the next, so the cleaning procedure must be developed on a case-by-case basis. The treatment needed to control bacteria in a relatively clean system also varies from one operation to another. Chlorine and sodium hypochlorite frequently have been found[50,52,53] to be very effective biocides, particularly in seawater. If the water contains large amounts of oxidizable substances or oxygen-scavenger chemicals, chlorine and hypochlorite may become uneconomical because of the amount wasted. Chlorine and hypochlorite may be brought in as a bulk product or generated on site by use of electrolysis to disassociate the chlorine from salt water. The size of the system and the logistics of the location will generally determine which is more economical.

Other effective biocides include amines, diamines, quaternary ammonium compounds, and formaldehyde. Recent studies[59,60,62] found that glutaraldehyde is particularly effective on slime-forming bacteria, algae, and SRB because of its ability to penetrate the biofilm and kill the bacteria within. Isothiazolone has been shown[60] to be effective against aerobic bacteria.

These biocides can be applied in batch or slug treatments periodically or added to the system continuously. There is a large variety of chemical pumps or feeders on the market that will inject chemicals into a water stream. The location of the injection point(s) is more critical than the device used for injection. Chemicals should enter the system in actively flowing water, not in dead spots. The location should allow adequate mixing time to give an effective kill. In some cases, the best procedure is to fill the system with the biocide and then to shut down, thus allowing the biocide to soak the system for some period of time. As discussed above, dead spaces, piping, or other areas with inadequate flow rates are poor design characteristics. Elimination of these will decrease corrosion problems.

There are many effective microbial-control systems that use various combinations of these chemicals and methods. For example, some North Sea waterfloods use continuous application of chlorine or hypochlorite and periodically batch treat with glutaralde-

hyde. In the Java Sea waterflood, hypochlorite is continuously injected with a glutaraldehyde/isothiazolone batch treatment given on a weekly basis.

The most effective treatment for a given water system will choose a chemical that is economical, compatible with the water and other chemicals used, nonplugging, and not so toxic as to cause handling problems. The engineer should consult water-treatment specialists to determine which chemical or combination will be optimum and how and where to apply the treatment. Both laboratory evaluation and field testing are essential to ensure that the proper choices are made.

4.4.6 Corrosion Control. One way to define corrosion is the interaction between a construction material and its environment. Obviously, there are two general ways to control corrosion. One is to choose or to modify the construction material to minimize corrosion. The other is to change the environment. Both these techniques will be discussed. Good design and correct operation of a waterflood system can eliminate or control the costly problem of corrosion. An example of how much corrosion can cost is found in the Horseshoe Gallup field near Farmington, NM. Oxygen was allowed to enter the water system at various points, and corrosion-inhibition treatments were stopped because of the cost. Subsequently, the injection-distribution system averaged more than 50 leaks per month. The cost of repairs was substantial. Similar experiences have occurred in many other fields where corrosion control was not adequate.

The primary corrosion problem in waterflooding is the interaction between steel and the water it contains. In controlling the environment, such methods as keeping air from contacting the water or deaerating water that has already been saturated with oxygen are applicable. Other methods discussed earlier, such as removing H_2S and CO_2 gases from the water and microbial control, also fall in the category of modifying the environment. Another method of this type is chemical treatment of the water with corrosion inhibitors. The most commonly used chemicals are the film-forming types, such as the amines and diamines. Quaternary ammonium compounds act similarly and have been successful in freshwater systems. These inhibitors are injected into the water stream at appropriate points and form protective films coating the steel in pipes, vessels, wells, and other parts of the system. The film protects the steel by isolating it from the water, thus preventing corrosion. The films are not permanent, so continuous or closely spaced periodic treatments with the inhibitor are required. The length of time the film will last is a function of the type of inhibitor used, flow velocity, temperature, and type of surface, among other factors.

Laboratory screening can be helpful in the selection of a chemical. However, trial-and-error testing in the field is required to determine rates and application method. Chemical supply companies are willing and able to do such testing on site when requested. It is recommended that several companies be asked to run these tests and to make proposals for treating. Comparison of the results should allow the engineer to choose the most economical alternative. This best choice may change through the treating system as the process removes oxygen, adds chemicals, or changes temperatures and pressures.

Cathodic protection is another way to modify the environment.[63] This technique uses sacrificial anodes and/or impressed currents to render the system entirely cathodic and thus prevent corrosion. Cathodic protection is most applicable to large areas of metal, such as the insides of vessels or tanks and the outsides of gathering- or injection-line systems.[64] Anodes will protect only the metal surfaces that are in their line of sight, as shown in Fig. 4.11.

The other approach to corrosion control is to change the structural material used. This method is best used in the initial design[65] of the systems to prevent costly retrofitting. Where appropriate and economical, materials such as fiberglass-reinforced plastic (FRP) or plastics of various types can be substituted for steel. Tanks and tubular goods made from these materials are readily available. For service such as low-pressure water-piping systems, high-density polyethylene pipe may be very economical and essentially eliminate any corrosion. Pipe made from FRP material has proved very effective in high-pressure injection service. Fiberglass lines should generally be buried to protect the line from the sunlight and for

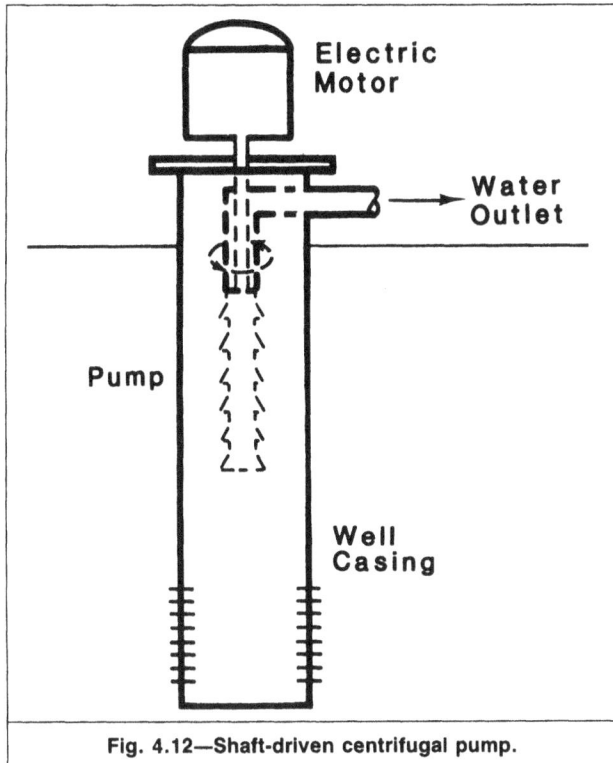

Fig. 4.12—Shaft-driven centrifugal pump.

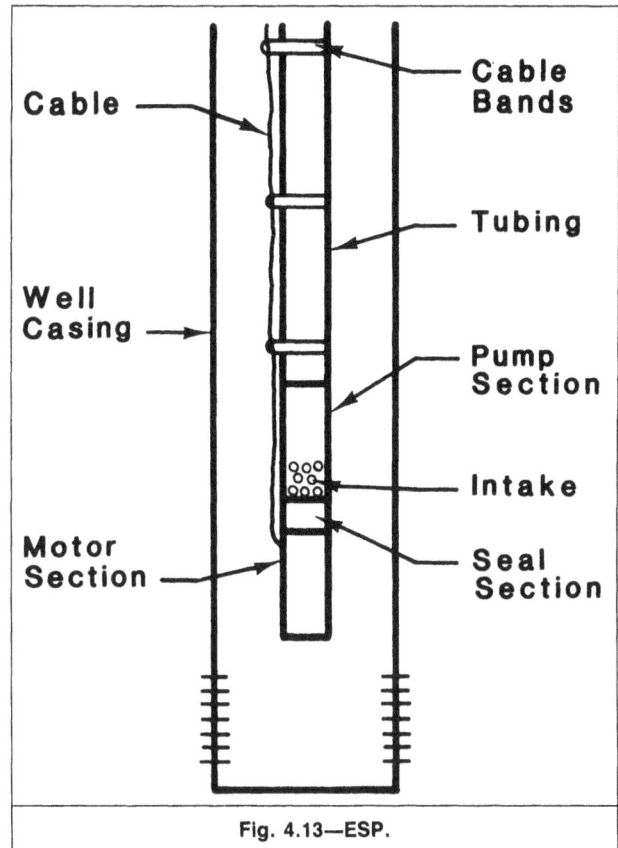

Fig. 4.13—ESP.

safety reasons. If FRP line is overpressured to the burst point, it will fail catastrophically and endanger nearby personnel. Pressure-relief systems and burial of the lines will minimize this problem. A safety concern with plastic or FRP vessels is the inability to ground the vessels against static electricity. Explosive vapors contained in the vessel could be ignited by a spark from the static charge.

A similar method to change the exposed materials is to line vessels[66] or lines with such coatings as cement or plastic. Many coatings are available, but they are all limited by temperature, chemical resistance, or other factors. The waterflood engineer should consult company standards and specialists in this field to pick a suitable coating for the service. Quality control in the coating applicator's procedures is absolutely critical in obtaining a good coating.[67] Buried steel pipelines justify the use of exterior coating and wrapping[68] along with cathodic protection.

The use of corrosion-resistant materials—such as stainless steel, aluminum-bronze, or more exotic metals—in valves, valve trim, pumps, and other smaller parts in the waterflood facility is recommended. The specification of these materials must be carefully written to ensure that the product will perform as expected. All stainless steels are not alike and must be chosen for the proper service and conditions.

Byars[67] concludes that "Properly planned, installed, and monitored corrosion control programs and proper material selection and specification are methods of failure control that will increase the profit for a project." The waterflood engineer is advised to keep that statement clearly in mind when designing a waterflood project.

4.5 Water Source Systems

4.5.1 Water Source Wells. Shallow water wells may be completed with slotted casing or a prepared screen opposite the water-bearing formation. A gravel pack can be used to prevent or to minimize sand production, but it may result in reduced productivity. Hydraulic fracturing is often essential, especially for deeper, saltwater sands. In general, source wells will have the same problems as oil-producing wells and the solutions are similar.

In shallow wells, shaft-driven pumps with the motor at surface, casing pumps, submersible pumps, or conventional oilwell-type pumping systems can be used. The shaft-driven pump, often referred to as a vertical turbine pump, is frequently used for shallow

water well installations < 1,000 ft [300 m] deep. This type of pump, shown in Fig. 4.12, consists of a vertical rotating shaft on which are mounted one or more impellers enclosed in a housing or bowl. The prime mover can be an electric motor mounted on top of the shaft at the surface or an engine connected to the shaft through a right-angle drive unit. The discharge head is located on the surface directly above the well.

According to McDuff and Pratos,[69] factors that must be determined for the selection of a pump are (1) the ID of the well, (2) the depth of the well, (3) the volume of water to be pumped per day, (4) analysis and specific gravity of the water, (5) prime mover type and speed of operation, (6) the static water level, (7) drawdown, (8) the well lift, which is the sum of the static water level and the drawdown, (9) the discharge height above ground plus pressure against which water is to be pumped, and (10) the depth at which impellers are set.

Conventional oilwell pumping equipment has been used to lift water, generally in volumes < 1,000 B/D [160 m³/d]. This is a fairly low volume for a water well, so oilwell pumping units are seldom used for water-supply wells.

Casing pumps have been used for water supply in a limited number of places. The main feature of the casing pump is the elimination of the tubing string. As the name implies, the pump works directly in the casing. It is operated with long, slow strokes. Volumes lifted may be up to about 3,500 B/D [556 m³/d] water in 4¾-in. [12-cm] casing from up to 2,000 ft [600 m] deep.[70]

Electrical submersible pumps (ESP's) are used in many applications at various depths. A submersible pump installation has three major downhole components: a motor, a pump, and a seal section. The seal section is attached to the motor, and above the seal is the pump itself, as shown in Fig. 4.13. To lower the equipment into the well, the tubing is screwed into the pumphead and the assembly is run into the hole with the three-conductor power cable banded to the tubing. The pump is a multistage centrifugal pump with the number of stages determined by the total lift required. Submersible pumps of this type can be very costly to run. Electrical failures in the cable, connections, or motor are not uncommon. It is essential that operating personnel are educated in the installation and operation of this equipment. It is important that the correct size of pump be used for each installation to prevent frequent on/off cy-

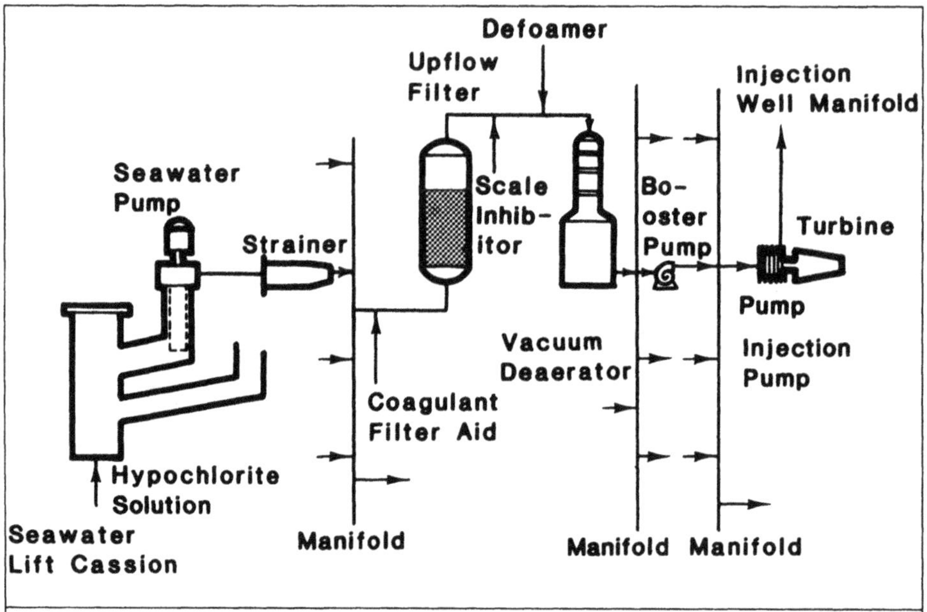

Fig. 4.14—Fateh waterflood system, Dubai (after Kennedy[71]).

cles, which reduce the lives of both the pump and motor. Variable-speed drive units can be purchased or rented to help size the pump and motor. There are submersible pumps with lifting capacities of 300 to more than 20,000 B/D [50 to 3200 m³/d] and head capacities of 12,000 ft [3660 m] or more. The same 10 factors listed as necessary information for pump selection, plus temperature at formation level, are required for selection of a submersible pump, motor, and cable. A submersible pump is frequently the only choice for a reasonable volume of water from the deeper wells. Careful study of the projected costs of each possible pump type should supply the correct choice for the installation. The initial, installation, and operating costs, including repairs and replacements based on experience factors, must be considered to arrive at the comparative costs per barrel of water produced.

4.5.2 Surface Water Sources. Ocean water sources are frequently produced by ESP's inside casings. Shaft-driven pumps have also been used successfully for seawater production. The casings are open to the seawater at a depth that yields the best water quality

as determined by sampling. The pump intakes will normally be set fairly near the surface. Shaft-driven pumps are also used for this service. Fig. 4.14 shows a schematic[71] of the Fateh field waterflood system in Dubai, which uses this type of pump. Good design practices suggest the installation of two or three pumps with spare capacity to allow a continuous water supply rate during repairs or maintenance. Another method of collecting surface waters is to build a reservoir or intake structure from which water is pumped with direct-coupled, vertical turbine ("canned"), or conventional centrifugal pumps. The waterfloods on the North Slope of Alaska use this method for source-water supply. These two techniques can also be applied to such freshwater sources as lakes and streams. At the Signal Hill East Unit waterflood, the source was fresh water supplied by a municipal system. The produced- and source-water systems for this flood are shown in Fig. 4.15.[72]

4.5.3 Source-Water Gathering Systems. The design of piping systems, including pressure-drop calculations, line-size-vs.-installed-power tradeoff, and layouts will be covered in Chap. 6. The con-

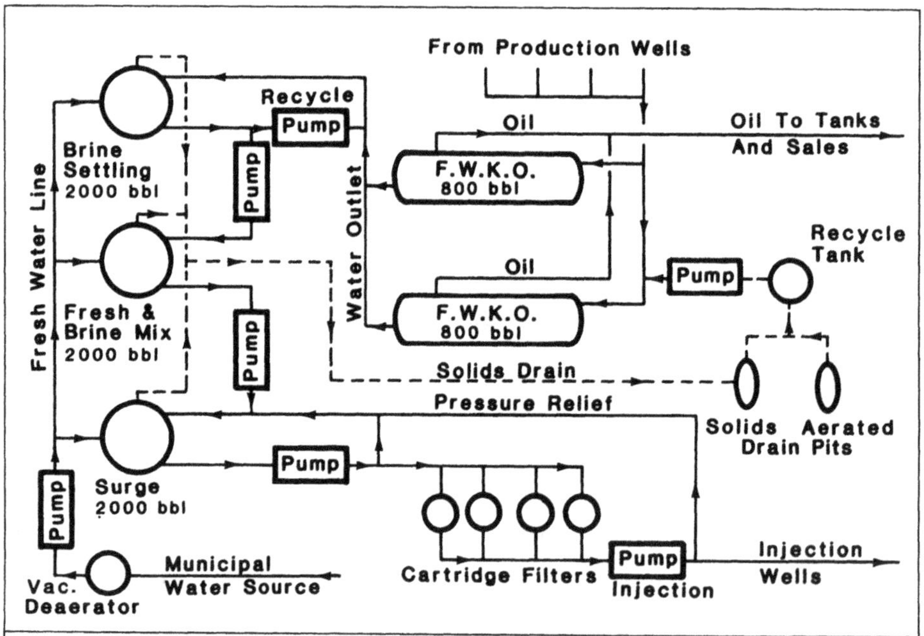

Fig. 4.15—Signal Hill East Unit waterflood system.

siderations discussed in Chap. 6 are equally applicable to the design of source-water gathering systems. These facilities differ from injection lines mainly in being at much lower pressures. Hence, plastic or FRP lines are more often used for source water than injection systems. Use of these types of materials for the lines is usually more economical than steel lines, particularly because no corrosion treatment is required to protect the lines.

References

1. Byars, H.G. and Gallop, B.R.: "Injection Water + Oxygen = Corrosion and/or Well Plugging Solids," paper SPE 4253 presented at the 1972 SPE Symposium on the Handling of Oilfield Waters, Los Angeles, Dec. 4-5.
2. Gates, G.L. and Parent, C.F.: "Water Quality Control Presents Challenge in Giant Wilmington Field," Oil & Gas J. (Aug. 16, 1976) 115-26.
3. "Recommended Practice for Analysis of Oil-Field Waters," RP-45, second edition, API, Dallas (July 1981).
4. Byars, H.G. and McDonald, J.P.: "How to Plan, Design and Operate Economic Water Injection Systems," World Oil (April 1968) 113-26.
5. Ostroff, A.G.: "Compatibility of Waters for Secondary Recovery," Prod. Monthly (March 1963) 2-9.
6. Bilhartz, H.L.: "Are We Making Water Systems Too Complex?" Waterflooding, Reprint Series, SPE, Richardson, TX (1959) 2, 33-36.
7. Hewitt, C.H.: "Analytical Techniques for Recognizing Water-Sensitive Reservoir Rocks," JPT (Aug. 1963) 813-18.
8. Jones, F.O. Jr.: "Influence of Chemical Composition of Water on Clay Blocking of Permeability," JPT (April 1964) 441-46; Trans., AIME, 231.
9. Veley, C.D.: "How Hydrolyzable Metal Ions React With Clays To Control Formation Water Sensitivity," JPT (Sept. 1969) 1111-18.
10. Reed, M.G.: "Stabilization of Formation Clays With Hydroxy-Aluminum Solutions," JPT (July 1972) 860-64; Trans., AIME, 253.
11. Barkman, J.H. et al.: "An Oil-Coating Process To Stabilize Clays in Fresh Waterflooding Operations," JPT (Sept. 1975) 1053-59.
12. Clementz, D.M.: "Clay Stabilization in Sandstones Through Absorption of Petroleum Heavy Ends," JPT (Sept. 1977) 1061-66.
13. Sydansk, R.D.: "Stabilizing Clays With Potassium Hydroxide," JPT (Aug. 1984) 1366-74.
14. Doscher, J.M. and Weber, L.: "The Use of the Membrane Filter in Determining Quality of Water for Subsurface Injection," Drill. & Prod. Prac., API (1958) 168-79.
15. Johnson, K.H. and Castagno, J.L.: "Evaluation by Filter Methods of the Quality of Waters Injected in Waterfloods," Report of Investigations 6426, U.S. Bureau of Mines (1964).
16. "Methods for Determining Water Quality for Subsurface Injection Using Membrane Filters," TM-01-73, Natl. Assn. of Corrosion Engineers, Houston (1976).
17. Barkman, J.H. and Davidson, D.H.: "Measuring Water Quality and Predicting Well Impairment," JPT (July 1972) 865-73; Trans., AIME, 253.
18. Cerini, W.F.: "How to Test the Quality of Injection Water," World Oil (Aug. 1958) 189.
19. McCune, C.C.: "On-site Testing To Define Injection-Water Quality Requirements," JPT (Jan. 1977) 17-24.
20. Maharidge, R.L.: "A Particle Sizing Technique Using Rayleigh Back-scattered Ultrasonic Waves," paper SPE 13643 presented at the 1985 SPE California Regional Meeting, Bakersfield, March 27-29.
21. "Preparation and Installation of Corrosion Coupons and Interpretation of Test Data in Oil Production Practice," RP-07-75, Natl. Assn. of Corrosion Engineers, Houston (1976).
22. Watkins, J.W.: "Design and Operation of Plants for the Preparation of Water for Injection into Oil Reservoirs," Waterflooding, Reprint Series, SPE, Richardson, TX (1959) 2, 15-21.
23. "Corrosion Monitoring System Provides Fast Measurement," Chilton's Instruments and Control Systems (Aug. 1983) 53.
24. Stiff, H.A. Jr.: "Corrosion: Holes in the Oil Industry's Pockets," Pet. Eng. (May 1959) B-19-B-24.
25. Ostroff, A.G.: Introduction to Oilfield Water Technology, Prentice-Hall Inc., Englewood Cliffs, NJ (1965).
26. Patton, C.C.: Oilfield Water Systems, Campbell Petroleum Series, Norman, OK (1974).
27. Battle, J.L.: "Salt Water Disposal," Proc., 15th Annual Southwestern Petroleum Short Course, Lubbock, TX (1968) 202.
28. Perry, L.N. and Frank, W.J.: "Sulfur Burning Method of Scavenging Oxygen from Water," paper 906-11-E presented at the 1966 API Southwestern Dist. Spring Meeting, Hobbs, NM, March.
29. Adams, G.H.: "Vacuum Deaeration in Waterflood Operations," Drill. & Prod. Prac., API (1968) 159-63.
30. Carlberg, B.L.: "Vacuum Deaeration—A New Unit Operation for Waterflood Treating Plants," paper SPE 6096 presented at the 1976 SPE Annual Technical Conference and Exhibition, New Orleans, Oct. 3-6.
31. "Design and Operation of Stripping Columns For Removal of Oxygen from Water," RP-02-78, Natl. Assoc. of Corrosion Engineers, Houston (1978).
32. Smith, C.R.: Mechanics of Secondary Oil Recovery, Reinhold Publishing Corp., New York City (1966) 250.
33. Weeter, R.F.: "Desorption of Oxygen From Water Using Natural Gas for Countercurrent Stripping," JPT (May 1965) 515-20.
34. Smith, J.P.: "Commissions and Operational Experiences of Java Seawater Treating Facility," JPT (July 1985) 1276-84.
35. Kissel, C.L. et al.: "Factors Contributing to the Ability of Acrolein To Scavenge Corrosive Hydrogen Sulfide," SPEJ (Oct. 1985) 647-55.
36. Weeter, R.F.: "A Practical Solution to Preparing a Sour Water for Injection," paper SPE 582 presented at the 1963 SPE Permian Basin Oil Recovery Conference, Midland (May 9-10).
37. Frank, W.J.: "Corrosion Control in a Waterflood by Removal of Hydrogen Sulfide and Carbon Dioxide from the Injected Water by a Hydrocarbon Gas Cycling Process," Proc., 14th Annual Southwestern Petroleum Short Course, Lubbock, TX (1967) 177-83.
38. Mitchell, R.W. and Finch, E.M.: "Water Quality Aspects of North Sea Injection Water," JPT (June 1981) 1141-52.
39. Zemel, B. and Bowman, R.W.: "Residence Time Distribution in Gravity Oil-Water Separations," JPT (Feb. 1978) 275-82.
40. Manual on Disposal of Refinery Wastes—Volume on Liquid Waste, first edition, API, New York City (1969).
41. Sport, M.C.: "Design and Operation of Dissolved-Gas Flotation Equipment for the Treatment of Oilfield Produced Brines," JPT (Aug. 1970) 918-20.
42. Degner, V.R.: "Dispersed Air Flotation Cell Design and Operation," Water, Symposium Series, AIChE, New York City (1975) 71, No. 151, 257-64.
43. Stickland, W.T. Jr.: "Laboratory Results of Cleaning Produced Water by Gas Flotation," SPEJ (June 1980) 175-90.
44. Leech, C.A. III et al.: "Performance Evaluation of Induced Gas Flotation Machine Through Mathematical Modeling," JPT (Jan. 1980) 48-58.
45. Bradley, B.W.: "Oil Field Water Handling—2," Oil & Gas J. (Nov. 18, 1985) 136-40.
46. Bradley, B.W.: "Oil Field Water Handling—3," Oil & Gas J. (Dec. 9, 1985) 42-45.
47. Caudle, D.D.: "Water Treating Plant Design and Operation," paper SPE 10006 presented at the 1982 SPE Intl. Petroleum Exhibition and Technical Symposium, Beijing, March 18-26.
48. Al-Arfaj, K.A. and Nomitsu, T.: "Waste Water Treatment Facilities and Disposal Well Injection System," paper SPE 9580 presented at the 1981 SPE Middle East Oil Technical Conference, Bahrain, March 9-12.
49. McCune, C.C.: "Seawater Injection Experience—An Overview," JPT (Oct. 1982) 2265-70.
50. King, P.J. and Robinson, K.: "Logical Approach Yields Correct Injection Water Quality," Pet. Eng. (Oct. 1981) 108-22.
51. Evers, R.H.: "Mixed Media Filtration of Oily Waste Waters," JPT (Feb. 1975) 157-63.
52. Jorque, M.A.: "How to Treat Seawater for Water Injection," Pet. Eng. (Nov. 1984) 28-34.
53. Mitchell, R.W. and Bowyer, P.M.: "Water Injection Methods," paper SPE 10028 presented at the 1982 SPE Intl. Petroleum Exhibition and Technical Symposium, Beijing, March 18-26.
54. Stiff, H.A. Jr. and Davis, L.E.: "A Method for Predicting the Tendency of Oil Field Waters to Deposit Calcium Carbonate," Trans., AIME (1952) 195, 213-16.
55. Skillman, H.L. Jr., McDonald, J.P. Jr., and Stiff, H.A. Jr.: "A Simple, Accurate, Fast Method for Calculating Calcium Sulfate Solubility in Oil Field Brine," paper 906-14-I presented at the 1969 API Southwestern Dist. Spring Meeting, Lubbock, March 12-14.
56. Templeton, C.C.: "Solubility of Barium Sulfate in Sodium Chloride Solutions from 25° to 95°C," J. Chem. Eng. Data (Oct. 1960) 5, No. 4, 514-16.
57. "Recommended Practice for Biological Analysis of Subsurface Injection Waters," RP-38, third edition, API, Dallas (Dec. 1975).
58. Costerton, J.W., Geesey, G.G., and Cheng, K.J.: "How Bacteria Stick," Scientific American (Jan. 1978) 86-95.
59. Smith, J.P.: "Commissioning and Operational Experiences of Java Seawater Treating Facility," JPT (July 1985) 1276-84.
60. Ruseska, I. et al.: "Biocide Testing Against Corrosion-causing Oilfield Bacteria Helps Control Plugging," Oil & Gas J. (March 8, 1982) 253-64.
61. Chen, E.Y. and Chen, R.B.: "Monitoring Microbial Corrosion in Large Oilfield Water Systems," JPT (July 1984) 1171-76.

62. Eager, R.G.: "Glutaraldehyde—A Microbiocide for Oilfield Applications," paper presented at the 1981 NACE Middle East Corrosion Conference, Bahrain, Jan.

63. "Design, Installation, Operation and Maintenance of Internal Cathodic Protection Systems in Oil Treating Vessels," RP-05-75, Natl. Assn. of Corrosion Engineers, Houston (1975).

64. Calberg, B.L.: "Water Quality for Trouble-free Injection," *Pet. Eng.* (Aug. 1980) 86–101.

65. "Selection of Metallic Materials To Be Used in All Phases of Water Handling for Injection Into Oil Bearing Formations," RP-04-75, Natl. Assn. of Corrosion Engineers, Houston (April 1975).

66. "Method for Lining Lease Production Tanks with Coal Tar Epoxy," RP-03-72, Natl. Assn. of Corrosion Engineers, Houston (1972).

67. Byars, H.G.: "Corrosion Control Programs Improve Profits, Part 2," *Pet. Eng.* (Nov. 1985) 50–58.

68. "Control of External Corrosion on Underground or Submerged Metallic Piping Systems," RP-0-69, Natl. Assn. of Corrosion Engineers, Houston (1976).

69. McDuff, J.R. and Pratos, C.A.: "Operation of Deep Set Shaft-driven Water Supply Pumps," *Proc.*, Ninth Annual West Texas Oil Lifting Short Course, Lubbock, TX (1962) 35–37.

70. Young, H.W.: "Casing Pumps," *Proc.*, Ninth Annual West Texas Oil Lifting Short Course, Lubbock, TX (1962) 102–03.

71. Kennedy, J.L.: "Offshore Water-Injection System at Dubai Is Expanded," *Oil & Gas J.* (May 30, 1977) 85–89.

72. "Injection Water Treating Manual," Arco Chemical Co., Oilfield Div., Plano, TX (1979) 71.

SI Metric Conversion Factor

$$\text{bbl} \times 1.589\ 873 \qquad \text{E}-01 = \text{m}^3$$

Chapter 5
Injection-Station Design

5.1 Introduction

The heart of every waterflood system is the pump station. The waterflood engineer must select the correct equipment for the pumping system and auxiliary systems. Selection of equipment that is best suited for the specific task is the basis for an economical and satisfactory installation. The location of pump stations must also be carefully chosen for the most efficient operation. This chapter will discuss the principles of design for the equipment and location of pump stations. At numerous points in this chapter, examples are given that include data from specific, named manufacturers. These are strictly examples and do not imply endorsement, nor is the intent to exclude any other vendors or their equipment.

5.2 Selection of Injection Rates and Pressures

Chap. 3 discussed the collection of field data and the calculations necessary for a reservoir study. The results of that study should include the rate and pressure needed for water injection. However, adjustments are frequently necessary to achieve a practical solution for a given set of circumstances. Waterflood performance depends on the volume and location of the injected water. Hence, the reservoir study should recommend a pattern and specify rates and pressures expected by well or pattern.

5.2.1 Injection Rates. Injection rates must exceed reservoir withdrawals if the reservoir pressure is to increase. Depending on reservoir conditions, an oil bank may form ahead of the injected water even without excess injection. However, the oil bank will generally develop more rapidly and the waterflood response occur sooner when injection exceeds withdrawal rates.

So, in most cases, field injection rates should exceed the concurrent reservoir withdrawal rate of produced oil, gas, and water. Generally, the more depleted reservoirs with lower pressures will require more excess injection to achieve response in a reasonable length of time. Craig[1] addresses the effect of rate on volumetric sweep calculations, concluding that "it is impossible to make a general statement as to optimum flooding rate because of the wide range of rock and fluid characteristics in oil reservoirs . . . ." Buckwalter[2] recommends injecting as much water as is economical in terms of available equipment and water-source availability. Additionally, the pattern that has the highest ratio of injection wells to producing wells should result in waterflood response in the shortest time, assuming the water is available and other factors are constant.

Injection into a particular well is obviously controlled by the reservoir in the immediate area of the well. Using Darcy's equation for flow of an oil bank followed by a water bank, Craig[1] showed that the injection rate before interference is described by

$$i_w = \frac{7.07 \times 10^{-3} hk\Delta p}{\left(\dfrac{\mu_w}{k_{rw}} \ln \dfrac{r}{r_w} + \dfrac{\mu_o}{k_{ro}} \ln \dfrac{r_e}{r} \right)}, \quad \dots \dots \dots \dots \dots (5.1)$$

where

i_w = water injection rate,
h = formation thickness,

k = absolute permeability,
μ_w = water viscosity at reservoir conditions,
k_{rw} = relative permeability to water,
r = radius of the external boundary,
r_w = well radius,
μ_o = oil viscosity at reservoir conditions,
k_{ro} = relative permeability to oil,
r_e = outer radius of the oil bank, and
Δp = $p_e - p_w$

where p_e is the pressure at the external boundary and p_w is the bottomhole pressure (BHP) at the injection well.

From Eq. 5.1, the injection rate will decline at a constant Δp because r_e is increasing from the wellbore outward toward the oil bank from an adjacent injection well. Calculating the injection rate during the period after interference occurs but before fill-up is a trial-and-error process. Subsequent to achieving fill-up, the water-injection rate will approximately equal the total reservoir voidage (both measured under reservoir conditions).

Returning to Eq. 5.1, kh must be evaluated to calculate the injection rate. The most common sources of kh are pressure-buildup curves taken during the producing life. This parameter can also be derived from core data, but caution must be used so that the data are representative of reservoir conditions. Other information needed includes relative permeability curves, which are obtained from reservoir calculations. Use of these factors is routine for reservoir engineers and is discussed more thoroughly in reservoir engineering textbooks. As a last resort, a rule of thumb suggests water-injection rates of 0.5 to 1.0 B/D-acre-ft* of reservoir.

One special case needs mentioning. It is often justified from a reservoir engineering standpoint to start waterflood operations while the reservoir pressure is above or only slightly below the bubblepoint. Under these conditions, which approximate reservoir fill-up, there is little incentive to increase the reservoir pressure. In such a case, the injection rate should equal the reservoir withdrawal rates desired plus known or expected losses, such as to thief zones.

5.2.2 Bottomhole Injection Pressure. According to Eq. 5.1, a critical factor in injection-rate calculations is the difference in pressure between the bottomhole injection sandface and the reservoir. The bottomhole injection pressure will usually be determined by reservoir conditions. Normally, the upper limit for the sandface injection pressure will be the formation parting pressure (see Sec. 5.2.2.1).

A column of fresh water at 60°F exerts a gradient of about 0.43 psi/ft. Thus the BHP in a 5,000-ft well, ignoring temperature effects, is 0.43 × 5,000, or 2,150 psi. For other than fresh water, or to account for temperature, this gradient is multiplied by the specific gravity of water at the given temperature. The injection rate calculated with hydrostatic pressure as p_w when calculating Δp will determine whether more pressure is needed. If, at the extreme, $p_e = 2,150$ psi in the case of a 5,000-ft well, Δp would equal zero and there would be no injection. If p_e approached zero, as it might

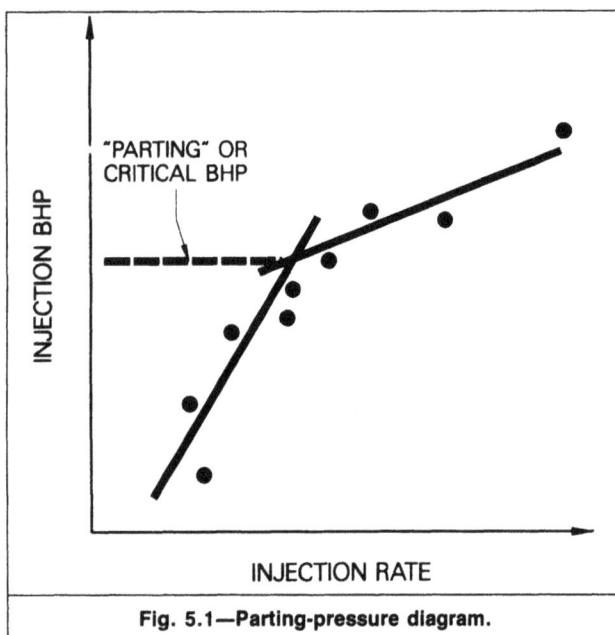

Fig. 5.1—Parting-pressure diagram.

in a nearly depleted field, then Δp will approach 2,150 psi, which may be more than sufficient to achieve the desired rates. If the well takes the needed amount of water in these conditions, it is said to "take water on a vacuum," meaning that no pressure is added to the hydrostatic head. A number of depleted fields take water on a vacuum. More commonly, the static head is not sufficient to inject the desired water quantity. Hydrostatic column pressure can be considered the lower limit of BHP.

5.2.2.1 Parting Pressure. During fluid injection, a critical pressure exists above which a well takes a higher fluid rate per increment of pressure increase. This is the result of a formation rupture or fracture and is called the "parting" or "fracturing" pressure. This parting pressure can normally be estimated by use of a gradient of 1.0 psi/ft times the well depth. A 5,000-ft well would then have a parting pressure of about 5,000 psi.

Actual parting pressures may vary[3] from 0.8 to 1.4 psi/ft. In many reservoirs, the parting pressure can be determined by pumping water into a well at increasing rates and pressures. When the injection rate vs. BHP is plotted, there will usually be a break in the curve, as shown in Fig. 5.1. The pressure at this break is called the parting pressure. Because no proppant is injected, the fracture will heal when the pressure is released. Because of the high leakoff rate of water, only shallow fractures will be formed.

Another source for estimating parting pressure is the breakdown pressure recorded during hydraulic fracturing treatments. This value is noted on the surface-pressure-vs.-time chart recorded by the fracturing company. The waterflood engineer must evaluate the friction loss so that an accurate BHP can be determined. Formation breakdowns at low rates give the best data for this purpose. Most fracture treatments also obtain an instantaneous shut-in pressure (ISIP). Howard and Fast[3] define this as the surface pressure (exclusive of friction losses) required to inject into the formation after breakdown. They say this will vary and can be as low as 0.57 psi/ft for an 8,000-ft well. The ISIP is not the parting pressure, but it is an indication of what happens to injectivity after a formation is parted.

Only rarely will the reservoir engineer allow injection pressures to exceed parting pressure. Therefore, the waterflood engineer should normally design for a BHP slightly less than the parting pressure. If no better data are available, the engineer can design for a maximum BHP equal to 1 psi/ft multiplied by the well depth. The well depth is the true vertical depth for a deviated well.

The waterflood engineer should discuss injection BHP with the reservoir engineer or evaluate it independently with standard reservoir engineering equations such as Craig's[1] shown above. The parting pressure will be the normal maximum for BHP. As stated above, high-permeability or depleted reservoirs may never need pressures approaching the parting pressure and may take water on

a vacuum. Consideration should be given, however, to designing the equipment to allow maximum pressures exceeding the parting limit in case formation blocking occurs.

5.2.2.2 Analogous Field Waterfloods. An often overlooked source for estimating injection BHP is a nearby waterflood in the same formation. Operators will usually share this information, at least to the extent of supplying wellhead pressures and rates. Most states require these data to be supplied on a monthly basis and they become a matter of public record.

Data from actual field experience tend to be more accurate than theoretical calculations. The engineer should get all available information on the operation to ensure that there are no problems that may have increased wellhead pressures or recent well stimulations that may have decreased wellhead pressures.

5.2.3 Wellhead Injection Pressure. The wellhead injection pressure can be determined by: surface pressure equals BHP minus hydrostatic head plus friction losses. Hydrostatic head can be calculated as outlined in Sec. 5.2.2. The friction loss in tubing is determined in the same manner as in other tubular goods. This is discussed in Sec. 6.1. Hence, once the injection BHP is determined, the wellhead injection pressure can be calculated with good accuracy.

5.2.4 Pump Discharge Pressure. Having converted a desired BHP to a required wellhead or surface pressure, the engineer must now determine a design injection-pump discharge pressure. At this stage of the calculations, simplistic approaches can be used. A trial-and-error solution can begin to determine the optimum economic choice of pump, driver, and injection-line system. A simple method is to assume a pressure loss from the pump discharge to the wellhead of 100 psi or 10% of the discharge pressure, whichever is larger.

As an example, assume that a BHP of 4,500 psi is chosen for a 5,000-ft well injecting 2,000 B/D water. The friction loss in the wellhead and downhole tubing is assumed to be about 50 psi. Then the surface pressure is $4,500 - (5,000 \times 0.43) + 50 = 2,400$ psi. So, a rough first estimate of pump discharge pressure for this case would be 2,400 psi divided by 0.9 to allow for 10% friction loss in the surface facilities, or 2,670 psi. Rounding off this value, the system design should be based on a pump discharge pressure of about 2,700 psi. Note that the design of the injection-distribution system was not specified except that a 300-psi total pressure drop was allowed in the system. Pumps and prime movers come in specified size ranges, so it is unlikely that a pump would operate at optimum efficiency at exactly 2,670 psi and 2,000 B/D water. The waterflood engineer must choose the best compromise from the available equipment. Equipment ratings such as horsepower and pressure must be balanced with available energy sources to achieve the best selection that meets the pressure and rate required. At this point, the engineer has defined a starting point where he knows approximate field injection rate and approximate pump discharge pressure. Given these factors, the engineer must consider which pump type is best and which type of prime mover will match the pump and available energy sources. This is not an exact science and there is not an optimum solution. The following sections will consider first the prime mover and then the pump selection.

5.3 Equipment Design Compromises

Now that pressure and rate requirements are estimated, the engineer faces the real world of equipment availability. Because no perfect solution is possible, compromise is necessary. As noted above, preliminary work will develop rounded numbers for the estimated rates and pressures, such as 4,000 B/D water and 1,500 psi. The data rarely justify more significant figures.

5.4 Prime Mover Evaluation

There are several standard types of prime movers: (1) natural-gas-fueled, internal-combustion engines, either naturally aspirated or turbocharged; (2) electric motors, usually AC but sometimes DC; (3) diesel- or gasoline-fired engines; and (4) natural-gas-fired turbines. Natural gas may be either pipeline-quality dry gas (mainly methane), wet casinghead gas, or a liquefied gas such as butane or propane.

In the selection of the best prime mover, a number of different factors should be considered.

1. Compatibility with existing equipment.

2. The equipment that field personnel are experienced with and accustomed to maintaining.

3. Availability of energy sources throughout the project life.

4. Size availability. Efficient turbines, for example, are not built in small sizes.

5. Energy consumption on an as-operated basis. Internal-combustion engines and electric motors operate on different efficiency-vs.-speed curves.

6. Cost of energy now and in the future.

7. Initial investment comparison between different types of drivers and between individual manufacturers.

8. Derating for existing conditions such as temperature and altitude. All units must be derated for high air temperatures and high altitudes, but by differing amounts. A gas turbine can be uprated for cold temperatures if the system is properly designed.

9. Compatibility with the driven unit (in this case, a water-injection pump). Rotating units, electric motors, or turbines fit well with rotating pumps, such as a centrifugal. It is easier to apply speed control to a fueled unit than to an electric motor.

The most important aspect is how well the horsepower and performance curves match the waterflood pump requirements. Obviously, a closely sized electric motor cannot meet a demand for an increased flow rate and head from the pump. Less obviously, an internal-combustion engine may have little or no reserve power if it is connected to a centrifugal pump. As discussed later, centrifugal-pump horsepower demand increases as a cubic function of speed. If a pump connected to a nominal 1,000-rev/min engine was originally sized for a 900-rev/min engine speed and then increased to require 990 rev/min, the 10% increase in engine speed results in a 33% increase in horsepower required by the pump. Because the developed horsepower of the engine is proportional to speed, the unit will put out only 10% more power. Unless the unit was considerably oversized to begin with, it is now undersized. When any variable-speed driver is connected to a centrifugal pump, the proper match must be carefully selected, allowing for future changes in conditions.

5.4.1 Horsepower Requirements.

The driver horsepower will be based on the pump and its mechanical and volumetric efficiency, the horsepower loss of the coupling system between the driver and pump, and the efficiency of the driver.

At this stage of the design, the engineer can only estimate the various efficiencies and horsepower losses. Reasonably close estimates can be made, however, because these factors vary only a few percent. Because the most common units in waterflooding are barrels of water injected per day and pounds per square inch, these will be used instead of gallons per minute and feet of head. Hydraulic horsepower, P_{hhp}, is the energy exerted on the water at the discharge flange of the pump and is expressed as

$$P_{hhp} = 1.701 \times 10^{-5} \times q_d \times p_d, \qquad \ldots \ldots \ldots \ldots \ldots (5.2)$$

where

P_{hhp} = hydraulic horsepower,

q_d = pump discharge flow rate, and

p_d = pump discharge pressure.

The input horsepower to a pump is stated in terms of brake horsepower, P_{bhp}. The input is greater than the output because of mechanical and volumetric efficiencies. These efficiencies are multiplied together to obtain total pump efficiency. Thus, a positive-displacement (PD) pump with a 5% volumetric efficiency loss and a 5% mechanical efficiency loss will be quoted as having a 90.2% efficiency. A typical centrifugal pump will have about 70% overall efficiency. The required brake horsepower input to the pump is then $P_{bhp} = P_{hhp} \times$ pump efficiency. Where the driver and the pump are coupled together by a mechanical device such as a gear box, horsepower losses will usually be increased by about 5%. The driver must be able to develop 105% of the calculated pump P_{bhp} to handle this intermediate loss.

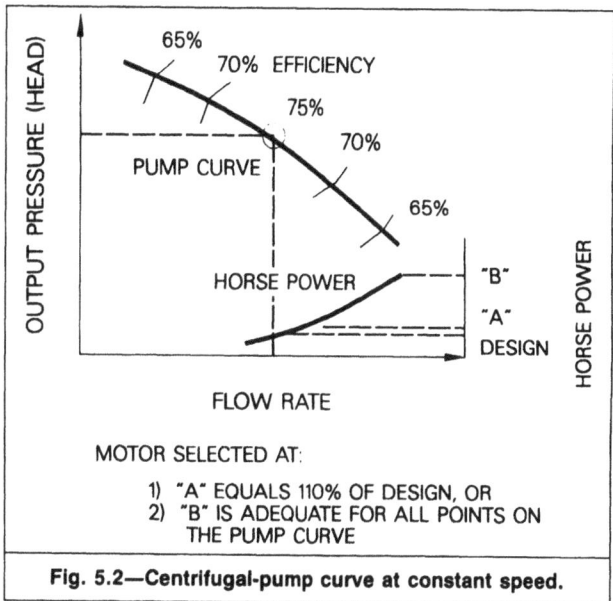

MOTOR SELECTED AT:

1) "A" EQUALS 110% OF DESIGN, OR
2) "B" IS ADEQUATE FOR ALL POINTS ON THE PUMP CURVE

Fig. 5.2—Centrifugal-pump curve at constant speed.

Although the details of the pump, connection, and driver are not determined at this stage, a convenient rule of thumb is to assume about 85% efficiency over the system. Then the equation becomes

$$P_{bhp} = 2 \times 10^{-5} \times q_d \times p_d = \frac{2 \times q_d \times p_d}{100,000}. \qquad \ldots \ldots \ldots \ldots \ldots (5.3)$$

From this, for example, 10,000 B/D water at 3,000 psi requires about 600 brake hp. Because the equation is roughly correct for a net efficiency of 85%, it will give close results for a PD pump but is too low for a centrifugal pump. But in this example, the design can start based on a total of about 600 brake hp in one, two, or more prime-mover/pump sets.

5.5 Electric Motors

5.5.1 Sizing Electric Motors.

If connected to a plunger pump, a motor should be sized to be able to drive the pump at the design horsepower plus any mechanical efficiency losses. Often, a larger motor is installed to match the maximum rated pump horsepower. This will result in excessive power bills while the motor is not loaded. Exactly how much will depend on electric rate structure and motor efficiency.

The design of electric motors is classified by the Natl. Electrical Manufacturers Assn. (NEMA) codes. The higher-rated motors will have better insulation and cores, resulting in more reserve built into the motor. Heat is the motor's enemy because it breaks down the insulation at high loads. But an oversized motor will be running at a point on its curve with penalties in terms of electrical efficiency. Generally, a good practice for a motor driving a PD pump is to select the next standard size motor above the pump frame rating with acceptable temperature rise and service factor. If, for some unique reason, the engineer chooses a smaller motor, the best motor of that size in terms of insulation and service factor should be installed.

For a motor-driven centrifugal pump, the problem is much more difficult. At constant speed, the centrifugal pump should be chosen so that it operates near the maximum efficiency point. The horsepower tends to increase at higher rates, even at a lower pressure, and tends to decrease at lower rates, even with higher pressures. At the extreme, horsepower approaches zero at zero rate, even though the pump discharge pressure is high. This is why a centrifugal pump is often started against a closed valve. If, for some reason, the pump is operated out on its curve—i.e., at higher rates—the motor will often overload and shut down.

A centrifugal pump seldom meets the engineer's precise design requirements. To be safe, the manufacturer will build the pump slightly above design to meet guarantees. Thus the horsepower required might be higher than expected. A common solution is to determine the motor horsepower required as some factor, such as

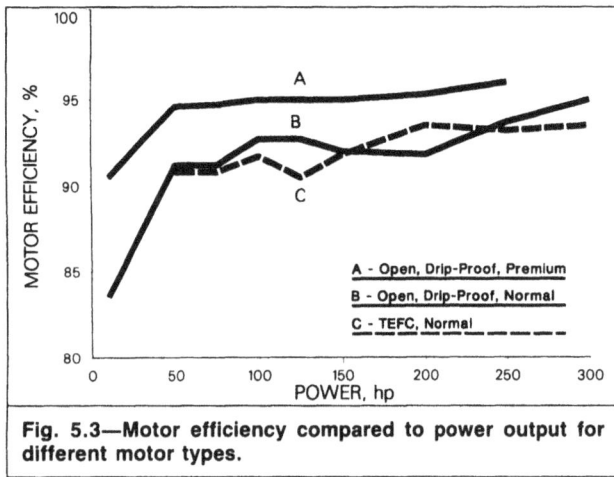

Fig. 5.3—Motor efficiency compared to power output for different motor types.

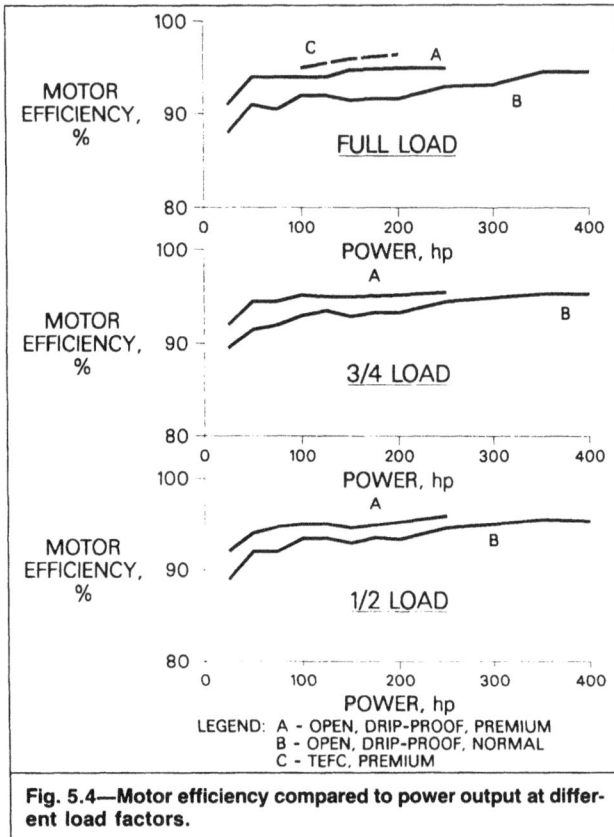

Fig. 5.4—Motor efficiency compared to power output at different load factors.

1.1, multiplied by the pump horsepower needed at the design rate (see Fig. 5.2). A more conservative (and safer) method is to select a motor that can operate the pump over the entire performance curve without overloading (Fig. 5.2). A pump must not be operated to the right of the vendor's published curve. The electric motor chosen should be the next larger standard size above the calculated horsepower requirements. If the 10% increase as shown by "A" in Fig. 5.2 jumps over a standard size, the situation should be evaluated with the help of the motor manufacturer with consideration given to expected life and energy costs.

Although the above is common practice and yields a conservative design, a more economical practice may be to consider a motor that closely matches the pump horsepower needs. If this is a standard motor size, then the purchase of that horsepower motor with a higher rated service factor and better insulation may be the best choice. A motor manufacturer's representative should be consulted for the best package on a near-100%-loaded design.

5.5.2 Motor Efficiency. All electric motors should be evaluated on efficiency, particularly the larger sizes. There are two efficiencies

to consider. One is for the unit running at rated horsepower. As shown in Fig. 5.3, there is a definite difference in efficiency, depending on the motor construction. In general terms, better mechanical construction gives a higher operating efficiency. This efficiency is purchased at an incremental cost that may pay for itself in reduced energy bills.

Unfortunately, the use of "efficiency" is not the same for all vendors. Three kinds of efficiency terms in common use are nominal, guaranteed, and minimum. Nominal efficiency is the usual value shown in brochures. A guaranteed efficiency requires that the manufacturer do a performance test and recoup his money for motors that fail the test. Thus, the purchaser pays for guaranteed efficiency and must consider the extra cost. The engineer is cautioned to determine exactly what efficiency a manufacturer is reporting.

The second efficiency factor is for the motor running at <100% rated horsepower. As shown in Fig. 5.4, motor efficiency varies only slightly over the horsepower range for a given load factor. The lowest efficiency generally occurs at full load, increasing as much as 2% as the load decreases to 75 or 50% rated load for a given horsepower. The engineer may be able to justify a larger motor by operating at lower loads but higher efficiency. For example, Fig. 5.4 shows 91.0% efficiency for a fully loaded, 75-hp motor but 92.5% efficiency for a 75%-loaded, 100-hp motor. The engineer would have to evaluate the 1.5% efficiency improvement against the extra cost of the larger motor.

An additional cost can be incurred when the power factor drops on partially loaded motors. With large motors, this decrease could trigger a penalty clause in the electric contract. On typical contracts, the demand charge is increased by about 1% for each 1% the power factor lags below 90%. Because demand makes up a significant part of the total bill, the electric cost could be increased drastically. Power factors can be corrected by installing capacitors into the circuit at an increased cost.

AC electric motors operating on a frequency of 60 cycles/sec have standard fixed speeds of 900, 1,800, and 3,600 rev/min. In most horsepower ranges, the 1,800-rev/min motor will be the least expensive, while slower speeds will be the most expensive. If the pump will operate at one of these fixed speeds, the motor can be coupled directly to the pump. This eliminates the need for a mechanical speed changer and the accompanying efficiency loss. Electronic equipment is now available to vary the speed of a motor over fairly wide ranges. These variable-speed drives (VSD's) may be worth installing in some cases. The subject is beyond the scope of this monograph, but manufacturers of VSD's and motors may be consulted. The more common method for speed adjustment is coupling the pump and motor through mechanical devices, such as belt drives or gear boxes. A gear-box speed ratio cannot be easily changed, but a belt-drive ratio can be adjusted by changing sheave size. Both systems have a 3 to 5% mechanical efficiency loss, which must be supplied by the motor output.

In summary, an electric motor should be selected to meet the following criteria.

1. It must be capable of either driving the greatest power requirement of the driven pump plus efficiency losses or of driving the pump at the desired rate plus losses. The latter eliminates any flexibility to handle future changes requiring higher horsepower.

2. The motor construction should be the best the engineer can justify in terms of insulation and efficiency.

3. A fixed-speed motor complementing the pump and a coupling mechanism should be used to give the least cost over the life of the system.

5.5.3 Motor Types. Waterflood equipment, by definition, operates in potentially wet locations. The waterflood engineer must select a motor for this environment. Three standard motor enclosure classifications exist.

1. Drip proof. This is an open motor that is dependent on a direct flow of air through the windings for cooling, for use where the ambient air is clean and dry with no airborne liquids. Some vendors have a "weather-protected" version in which the housing is protected against rain but not against a pressured stream of water.

2. Totally enclosed fan cooled (TEFC). In these motors, the winding is completely enclosed by the case and is cooled with air from an external, integrally mounted fan. This type can be used inside

or outside and will resist modest amounts of moisture up to occasional "baths." Most vendors have special versions for more severe service.

3. Totally enclosed/explosion proof. These motors have an enclosure designed to contain an explosion and to prevent its propagation to the surrounding atmosphere. These are for use in an atmosphere that may contain flammable gases. When these gases are present in the atmosphere most of the time, the Natl. Electric Code service classification is Group D, Class I, Div. 1. When the area carries oil or gas in piping only, the classification is usually Class I, Div. 2, with less severe restrictions. The Natl. Electric Code should be reviewed if hydrocarbons (oil or gas) are present in the area.

In virtually all cases, a TEFC motor is most suitable for waterflood installations. The probability of loss of an open motor is too high to be acceptable. Most waterflood installations will be classed in Div. 2 or even in the nonhazardous category. If the pump is in a hazardous area, serious consideration should be given to building a separate facility. If this is not possible (for example, on an offshore platform), then explosion-proof enclosures should be specified.

5.5.4 Auxiliary Equipment. Auxiliary electrical equipment is needed for electric motor operation. All motors require motor control equipment, which ranges from a simple on/off switch and a breaker for small motors to a sophisticated control center for larger motors. Included in this equipment should be devices to protect the motor against lightning, overvoltage, overload, short circuits, and undervoltage. Particularly in humid climates, an accessory to consider is a space heater to eliminate the condensation of water in the motor when it is not operating. The need for the other controls should be evaluated on a case-by-case basis.

The ability to start a loaded motor depends on both the motor design and the power supply. Often the power supply cannot start a large motor across the line—i.e., at full load. In these instances, complex starters must be installed to "soften" the starting load. These starters will be considerably more expensive than across-the-line units. Where large motors are used, the engineer should discuss the installation with the utility company or otherwise evaluate his power supply. "Large" is a subjective judgment, but any motor over 100 hp probably should be examined.

Power must be brought to the motor control and motor. This will normally require a transformer to convert from main-line voltage to motor voltage. Most oilfield motors will operate on 480-V, three-phase current but larger ones (600 hp plus) may be installed at less cost and operate more satisfactorily at 2,300 or 4,160 V.

Electric motor installations are complex enough to require discussion between the waterflood engineer, expert personnel, and vendors even for relatively small installations. For large motors or more complicated systems, the engineer should consult with experienced electrical engineers for design help. Capacitors should be considered for all motors larger than about 20 hp to maintain a high power factor in the distribution system. The most common installation switches the required capacitor on and off with the motor. This is especially advantageous where the operator owns the secondary distribution system. For a waterflood station, the best choice may be a control capacitor bank to handle all motors. Again, except for simple on/off-switched capacitors on small motors, experienced personnel and the vendor should be consulted on installation.

5.5.5 Effect of Nonstandard Conditions. Standard motors are built for <104°F ambient temperature at altitudes below 3,300 ft. As the ambient temperature increases, the insulation life decreases by half for each 18°F increment in operating temperature. Because motor designs are based on temperature rise above ambient, a higher air temperature results in a higher absolute operating temperature. It should be remembered that for a motor running inside a building, the ambient temperature is measured inside. The effect of increased temperature can be offset by purchasing a motor with better insulation that can tolerate the higher temperature. Below 3,300 ft altitude, Class B insulation can be used at 32 to 104°F; Class F at 104 to 149°F; and Class H at 149 to 185°F. These ranges allow for a temperature rise of at least 144°F in each case. Table 5.1 shows the allowable temperature rise above a 104°F ambient temperature.

TABLE 5.1—ALLOWABLE TEMPERATURE RISE

Motor Type	Allowable Temperature Rise Above 40°C Ambient* (°C)		
	Class B	Class F	Class H
Open, drip proof	80	105	125
TEFC	80	105	125

*Service factor = 1.0, altitude less than 3,300 ft.

The effect of altitude should be discussed with the vendor. The standard motor is designed for 32 to 104°F ambient from sea level to 3,300 ft altitude using Class B insulation. Class B insulation will tolerate a 144°F temperature rise for TEFC motors with a service factor of 1.0. For evaluations of 3,300 to 9,900 ft, Class F insulation should be used; Class H should be used at higher elevations up to 12,000 ft.[4] Motors must also be derated for altitude. One vendor suggests a factor of 1/2% decrease for each 330-ft increase in elevation over 3,300 ft if the temperature is <104°F ambient.

Motors are usually rated at service factors of 1.0 and 1.15, although higher factors are available. A service factor of 1.0 indicates that the motor can operate continuously at the rated horsepower without exceeding the temperature limits of the winding. A factor of 1.15 indicates that the motor can operate at 15% overload continuously without damage.

5.5.6 Codes and Standards. Some of the codes and standards that must be used for electrical system installations are listed below. In addition, local and state codes must be consulted to ensure compliance.

1. Natl. Electrical Code, latest version. Available from most electrical equipment vendors or from the Natl. Fire Protection Assn., 470 Atlantic Ave., Boston, MA 02210.

2. NEMA standards (for motors). Available from NEMA, 2101 "L" Street NW, Washington, DC 20037.

3. Instrument Soc. of America standards for controls. Instrument Soc. of America, 67 Alexander Dr., P.O. Box 12277, Research Triangle Park, NC 27709.

4. API RP 500A-C and RP 540-541 and API Standards 542 through 544. API, Production Dept., 211 N. Ervay, Suite 1700, Dallas, TX 75201.

5.6 Multicylinder Internal-Combustion Gas Engines

5.6.1 Gas Engine Sizing. Industrial engines are rated against various standards. The potential buyer should be fully aware of the differences because an operating horsepower specified by one vendor is not necessarily the same for another manufacturer. The standards usually encountered are set by the following groups: Diesel Engine Manufacturers Assn. (DEMA), API, Internal Combustion Engine Inst., and Soc. of Automotive Engineers. DEMA standards are not limited to diesel fueled engines. All specified horsepowers are based on conditions of near sea level at an air temperature usually not to exceed 90°F.

DEMA[5] defines an engine as complete with all essential equipment, piping (except that required for external cooling of jacket water), fans, and external lube-oil cooling pumps. The DEMA ratings also do not include losses from such items as air intake filters or exhaust mufflers. When evaluating engines rated on the DEMA system, the engineer is advised to request a performance curve with the fan, water-cooling pumps, exhaust muffler, and other required items in place.

API[6] defines an engine much differently in Standard 7B-11C. The "bare" unit includes the mounted water pump but no fan. A "power unit" includes both the fan and special water pumps. Radiators are available with fans that draw 1.5 to 2.5% of the engine rating, but other combinations use 5 to 6%. These percentages are calculated at full load. For an engine at 20% load, the low-draw system increases fuel consumption 6 to 8% compared with 12 to 16% with the high-draw unit. A fan/radiator system must have adequate cooling at peak power demand and maximum design ambient temperature. Cooling requirements change drastically with variations in conditions. For example, a system designed for 100% fan

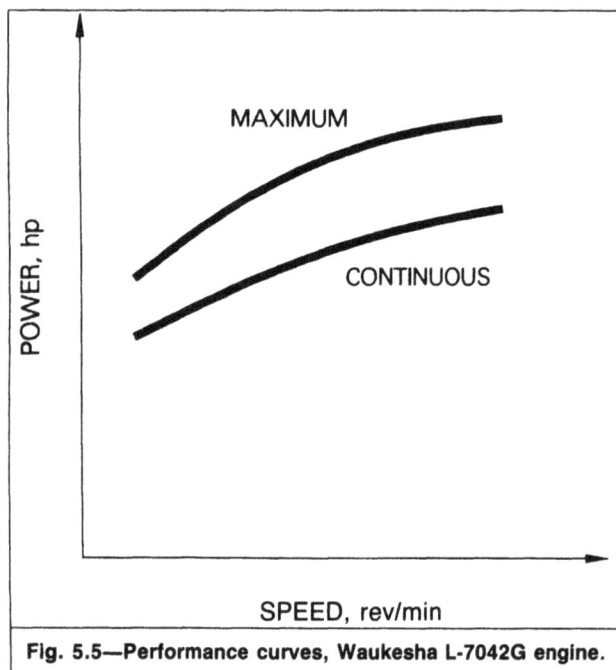

Fig. 5.5—Performance curves, Waukesha L-7042G engine.

TABLE 5.2—API POWER RATING

Unit	L7042 G (naturally aspirated)
Fuel	Gas
Maximum bhp at 1,200 rev/min	1,280
Intermittent bhp at 1,200 rev/min	1,152
Continuous bhp at 1,200 rev/min	1,024

TABLE 5.3—DERATING FACTORS

Organization	Rating Condition Limits	Percent Derate Over the Limit	
		Per 1,000 ft	Per 10°F
Naturally aspirated			
DEMA	1,500 ft and 90°F	3.5	2.5
API	500 ft and 85°F	3.0	1.0
Caterpillar	500 ft and 77°F	3.0	1.0
Waukesha	2,000 ft and 100°F	2.0	1.0
Turbocharged			
Caterpillar	6,500 ft and 65°F	3.0	1.0
Waukesha	3,000 ft and 100°F	2.0	1.0
White-Superior	4,000 ft and 100°F	3.5	2.5

TABLE 5.4—HYDROCARBON HEATING VALUES

	HHV or Gross (Btu/ft^3)	Air Required for Complete Combustion (ft^3/ft^3 hydrocarbon)	LHV or Net (Btu/ft^3)
Methane	1,009	9.53	909
Ethane	1,769	16.67	1,618
Propane	2,517	23.82	2,316
Butane	3,262	30.97	3,010
Pentane	4,008	39.98	3,707
Hexane	4,756	46.21	4,404

speed at 100% load with a temperature of 125°F requires only 50% fan speed at full load at 85°F. Conversely, the same half-speed fan is adequate for 125°F running at half load.[7] API 7B-11C defines maximum standard brake horsepower as follows: "At any rotational speed, the greatest horsepower, corrected to 85°F and 29.38 in. of mercury, that can be sustained continuously [based on a minimum of three readings taken at 5-min. intervals]."

The power curve for an engine is determined at a minimum of four evenly spaced speeds. A manufacturer must meet specific standard test conditions. The builder of an API-rated engine is required to furnish the purchaser (when requested) with curves of the maximum standard brake horsepower for both the bare- and power-unit engine. An "off-the-shelf" engine must develop at least 95% of the maximum brake horsepower shown on the curves to meet the standard. So, the basic brake horsepower for the API standard is 95% of the essentially instantaneous horsepower with a bare engine. It is obvious that this rating will differ from the DEMA standard.

Ratings for an example API-rated engine, the Waukesha, given in the 1981 *Diesel & Gas Turbine Worldwide Catalog* are shown in Table 5.2. The continuous brake horsepower is 80% of the maximum according to these data. The curves for this engine are shown in Fig. 5.5. Further reductions are necessary for fans and other auxiliary equipment, as well as temperature and altitude when appropriate.

The DEMA and API standards cover most engines except those made by Caterpillar Tractor Co. This manufacturer defines horsepower for its engines differently. Maximum is the horsepower that can be demonstrated within 5% for 5 minutes at the factory. Intermittent is the horsepower and speed capability that can be used for about 1 hour followed by 1 hour of operation at or below the continuous rating. Continuous is the horsepower and speed capability of the engine that can be used without interruption or load cycling. These definitions are at conditions of 29.61 in. Hg and 77°F. The engine is equipped with standard accessories, including air cleaner, lube-oil pump, fuel-oil pump, and jacket water pumps. Auxiliaries such as cooling fans, air compressors, and charging

alternators are not included. The engineer must clearly understand what is meant by horsepower and the effect of the accessories to evaluate the performance curves for various engines.

There are normally three speed ranges for engines: low speed is up to 300 rev/min and is not usually used in waterfloods, medium speed is 300 to 700 rev/min, and high speed is over 700 rev/min. The lower-speed units cost more per brake horsepower than the higher-speed engines and are usually heavier. Advocates of each speed range will expound on why that speed engine has the lowest overall lifetime cost. The engineer must evaluate these claims on the basis of the specific project conditions. Generally, the high-speed units wear out quicker but can be overhauled or replaced faster and less expensively. This compares with the long life of the low-speed unit with its attendant high-cost overhaul and longer out-of-service time during the overhaul.

5.6.2 Effect of Nonstandard Conditions. As mentioned above, the rated engine horsepower is for conditions of near sea-level altitude and for ambient temperatures below 90°F. Corrections must be made for a naturally aspirated unit operating at other conditions. Table 5.3 outlines the amount of derating necessary to correct for altitude and temperature. Both naturally aspirated and turbocharged engines are shown. If temperatures exceed the limits for only short periods of time occasionally during the year, it is not necessary to derate the engine. If the engine will be inside a building and using air from within the building, this must be included in the design because the air may be much warmer (or cooler) than the outside air.

Fuel characteristics can affect engine horsepower ratings. Gas engines run best on almost pure methane gas (usually referred to as pipeline quality) with a high heating value (HHV) near 1,000 Btu/ft^3 and a low heating value (LHV) of about 900 Btu/ft^3. The difference between the two numbers represents the heat lost in the exhaust with the water formed during combustion.

The engine will develop less horsepower as the heating value decreases, but the decrease is not linear. This power loss is the result of the inefficiency of handling inert gases. This inefficiency can be somewhat mitigated by using a turbocharged engine. Heating values above normal increase the available mechanical horsepower from an engine, but the increase should be used only after consulting the manufacturer to ensure that the engine will not be thermally overloaded.

Increased heating value comes from heavier hydrocarbons, as seen in Table 5.4.[8] The amount of air required for combustion also

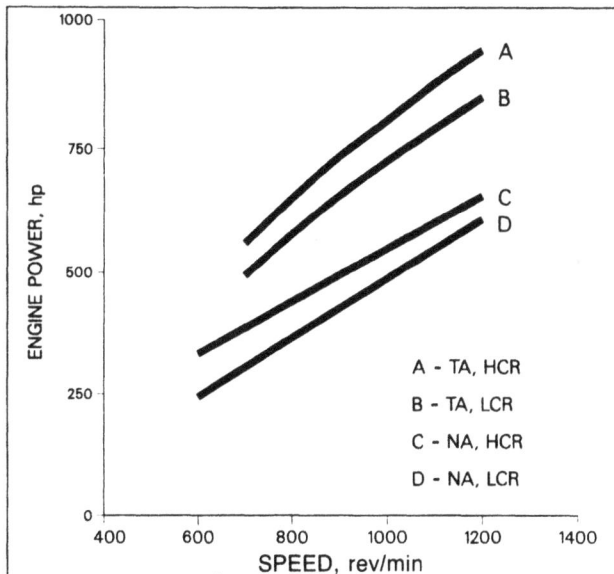

Fig. 5.6—Power curves for different versions of Caterpillar G-399.

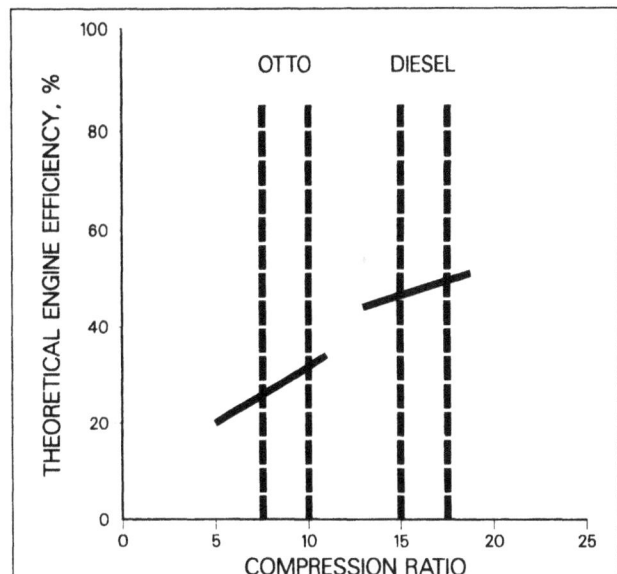

Fig. 5.7—Theoretical efficiency, Otto and diesel cycle engines.

increases from methane to hexane. Therefore, the actual fuel value is the heating value of the correct air/fuel mixture. This can be calculated by dividing the LHV of the combustible gas by the sum of the volume of the fuel gas (1 ft^3) and the amount of air needed to burn it. For example, methane's "true" heating value is 909 Btu/ft^3 (LHV) divided by (1.0 + 9.53) or 86.3 Btu/ft^3 of air/fuel mixture.

As the heating value of the gas increases, the ability of a high-compression-ratio (HCR) engine to burn the fuel without damage decreases. The vendor will usually base the criteria on butane or, possibly, propane content. Above a certain butane or propane percentage, a low-compression unit is the best choice. At even higher percentages, the best economic selection is limited to a naturally aspirated, low-compression-ratio (LCR) engine. Some engines can be set up to run on either methane or propane, but the manufacturer should be consulted to determine whether the engine must be derated.

If the engineer plans to run the engines on "wet field gas" (unprocessed and often not dehydrated gas), then the Btu value must be evaluated accordingly. If the engines in the design plan will not run properly on field gas, then another engine must be used or the wet gas must be processed. Gas processors remove the natural gas liquids from the gas stream wholly or partially down to the ethane level. Skid-mounted units are available, and the liquid produced will help pay for the initial and operating costs. The engineer must evaluate the economics of gas processing vs. changing the engine design.

5.6.3 Turbocharged Gas Engines. Table 5.3 shows the derating factors for naturally aspirated and turbocharged engines. Most manufacturers of intermediate and large engines sell turbocharged units. These units produce more horsepower with the same piston displacement as the naturally aspirated engines and can compensate for nonstandard altitude and temperature conditions. DEMA states that there is no simple formula for derating a turbocharged unit and that the engine manufacturers "will rate their engines in accordance with the compensating equipment furnished." API similarly states that the information concerning temperature and altitude conversion should be secured from the manufacturer. For certain vendors, this information is shown in Table 5.3. If the turbocharged unit also has intercooling, the unit must be additionally derated if the intercooler water temperature is above 85°F.

According to Moe,[7] an improper turbocharger match—i.e., running an engine at 1,000 rev/min when the unit was purchased for 1,200-rev/min full-load service—increases fuel consumption 1 to 5%. On the other hand, manufacturers can supply a more effective

turbocharger for high-altitude operation that can give fuel savings up to 2 to 4%.

Turbocharged engines with high continuous loading tend to incur higher maintenance and repair costs than equivalently loaded, naturally aspirated engines. The engineer should consult the literature and check on local facilities' experience to quantify the economics of turbocharged engines.

5.6.4 Compression Ratios. Most manufacturers sell engines with at least two different compression ratios. Although the actual ratios will vary, the resulting engine is usually classified either as HCR or LCR. These classifications apply to both naturally aspirated and turbocharged engines. LCR units are needed where only low-Btu gas is available. As mentioned above, LCR units are also the best choice for high-Btu gas. HCR engines are best used in the intermediate range of fuel-gas Btu value (approximately 900 to 1,100 Btu/ft^3). Fig. 5.6 shows the relationship of developed horsepower from four engines of a particular base type.[9] Given the same basic engine frame, the turbocharged engine (TA) will always have a higher power rating than the naturally aspirated (NA). For the units shown, at 1,000 rev/min, the TA engine provides 45 to 47% more power. Also, the HCR engine will develop more power than the same type of LCR engine. At 1,000 rev/min, the HCR engine develops 9 to 11% more power than the LCR unit for this particular engine.

Engine efficiency is directly related to compression ratio,[10] as shown in Fig. 5.7. As compression ratio decreases, the efficiency decreases, causing an increase in fuel consumption. A gas engine with the best currently available technology uses 7,100 Btu/hp-hr. The probable maximum efficiency achievable is 6,100 Btu/hp-hr and the ultimate theoretical is 4,200 Btu/hp-hr fuel usage. Note that these numbers do not include losses from auxiliary equipment.

5.6.5 Engine Performance. The operating characteristics of the four engines in Fig. 5.6—i.e., (A) TA-HCR, (B) TA-LCR, (C) NA-HCR, and (D) NA-LCR—will be looked at in more detail. It should be noted from Fig. 5.6 that developed power decreases from Unit A to Unit D at the same rev/min.

Fig. 5.8 compares the fuel consumption for these engines. It is obvious that Unit A, the TA-HCR, uses much less fuel per horsepower developed than Unit D, the NA-LCR. If 800 hp were needed and these engines were the only options, Unit A would use 800 hp×7,500 Btu/hp-hr or about 6,000 ft^3/hr of natural gas. If two engines like Unit D are chosen to develop the needed power, ignoring some difference in not being fully loaded, they would use 800 hp×8,450 Btu/hp-hr or 6,760 ft^3/hr of gas. Thus, choosing Unit D for power would use 12.6% more fuel than using Unit A.

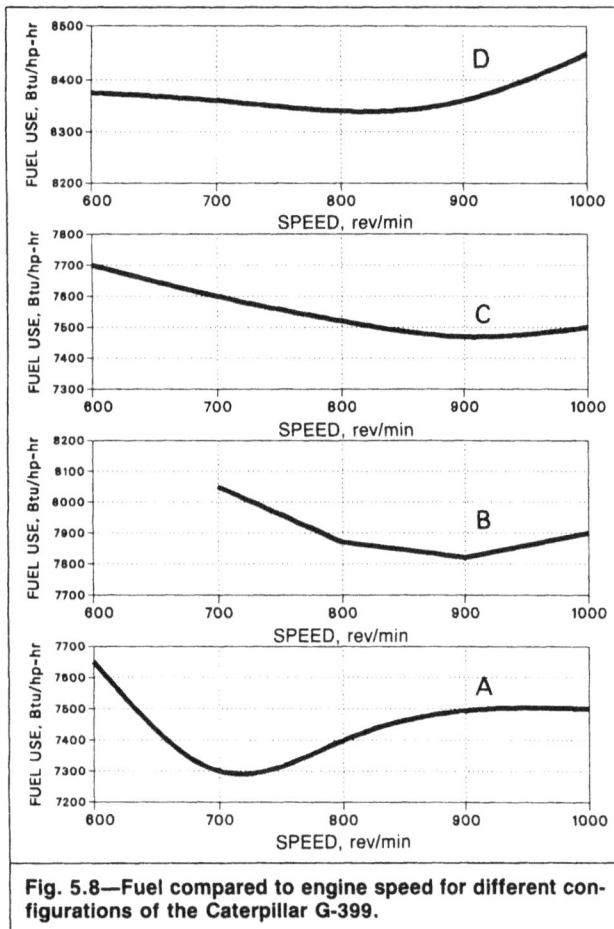

Fig. 5.8—Fuel compared to engine speed for different configurations of the Caterpillar G-399.

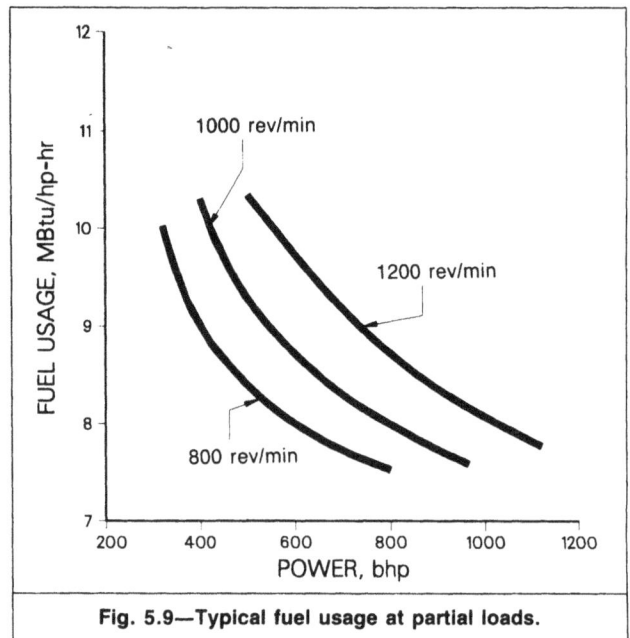

Fig. 5.9—Typical fuel usage at partial loads.

Now compare more similar engines, Units A and C; both are HCR but one is naturally aspirated and the other is turbocharged. They use the same fuel per brake horsepower-hour at 1,000 rev/min. The engineer must compare the engines on the basis of horsepower, installed cost, and maintenance costs, but not on fuel consumption. For example, if both units were operated at 500 hp, Unit A would operate at about 650 rev/min and consume about 7,550 Btu/bhp-hr, where Unit C would be fully loaded and use 7,500 Btu/bhp-hr. There is little difference in fuel consumption, but there is a large difference between the original investments and there may be significant differences in maintenance costs. If the initial horsepower demand was 500 hp but about 750 hp would be needed later in the waterflood, then Unit A is clearly the best choice among these four engines.

As has been discussed, centrifugal-pump horsepower requirements increase by a cube of the speed of the pump. Using the data from Fig. 5.6, the problem caused by this cube factor can be illustrated. If Unit A is loaded to 500 bhp operating at 650 rev/min, and later it is necessary to speed up the pump to an engine speed of 1,000 rev/min, the power needed is $(1,000/650)^3 \times 500$ or 1,820 bhp. But the engine is rated at only 800 hp at 1,000 rev/min. Alternatively, the pump load would have to be initially limited to 200 bhp at 650 rev/min engine speed. Operating an engine at 25% load is obviously a poor practice unless horsepower requirements are expected to increase early in the waterflood.

On the other hand, the horsepower ratio of 1.6 at 1,000 vs. 650 rev/min is about equal to the speed ratio of 1.53. So, a PD-type pump could be handled by the engine at either speed or any intermediate speeds. Fig. 5.9 from Marquis[10] shows the effect of partial loads on fuel consumption and, therefore, fuel economy.

Assuming a direct-drive PD pump and horsepower requirements of about 600 hp instead of the rated (by the curve) 960 hp, it is obvious that the load could be developed easily at 800 rev/min. As shown, the fuel consumption would be about 8,000 Btu/hp-hr at 800 rev/min instead of 9,400 Btu/hp-hr at 1,200 rev/min. Speeds

below 800 rev/min are normally unacceptable in a 1,200-rev/min engine.

The test stand horsepower discussed above is for a new engine. If the unit is fully loaded when new, it will not be able to carry the load after wear has occurred. Some additional derating should be considered so that the engine can drive the pump even after a reasonable amount of wear has taken place. Company standards or expert personnel should be consulted for the factor to be used. All engines eventually require major overhauls, but this technique can increase the time period between such jobs.

It is inherent that operating costs will increase substantially when an engine is run at maximum continuous load. The engineer should remember this when sizing an engine. As it wears, the engine will run closer and closer to its limit, and expenses will tend to rise asymptotically. Conservative design for horsepower needs will usually pay off in a waterflood with a long life. Design and planning require the "long view" of the project rather than minimization of initial costs. Many vendors do not have this viewpoint, so the engineer should approach their suggestions with caution.

5.6.6 Gas Engine Summary. Recapping the preceding sections on medium- to high-speed, multicylinder engines, the following statements can be made.

1. Gas-fueled, internal-combustion engines driving centrifugal pumps must be very carefully sized to provide sufficient horsepower for future increased injection rates.

2. Turbocharged engines have more rated horsepower than naturally aspirated engines of the same style. Although not discussed above, the resultant installed cost per brake horsepower is lower with a turbocharged unit. Turbocharged engines may have higher maintenance costs.

3. HCR engines produce more horsepower than LCR units. If the fuel gas meets the HCR requirements, the HCR engine will cost less per brake horsepower developed than a LCR unit.

4. Changing between naturally aspirated and turbocharged or between HCR and LCR will usually require a change in the base engine, although, in some cases, kits are available to do this. Therefore, the engineer must closely evaluate the options available during the initial design. For example, if gas with a low Btu value is available locally at low prices because of a lack of market, whereas high-Btu gas is available only at consumer cost, then the naturally aspirated, LCR engines may have an economic advantage.

5. The engineer must carefully evaluate basic horsepower definition, piston speed (wear is a direct function), derating factors for altitude and temperature, HCR vs. LCR (with attendant restrictions),

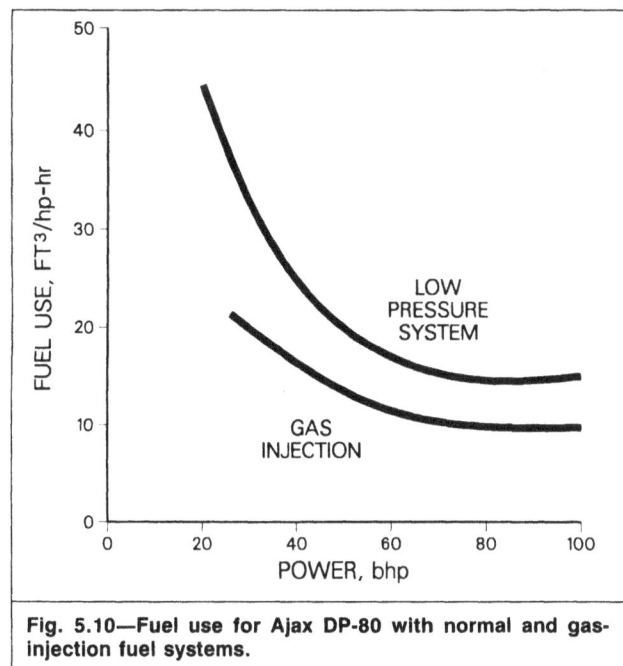

Fig. 5.10—Fuel use for Ajax DP-80 with normal and gas-injection fuel systems.

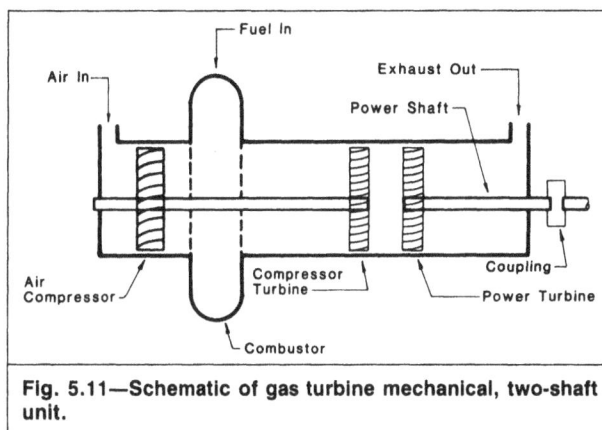

Fig. 5.11—Schematic of gas turbine mechanical, two-shaft unit.

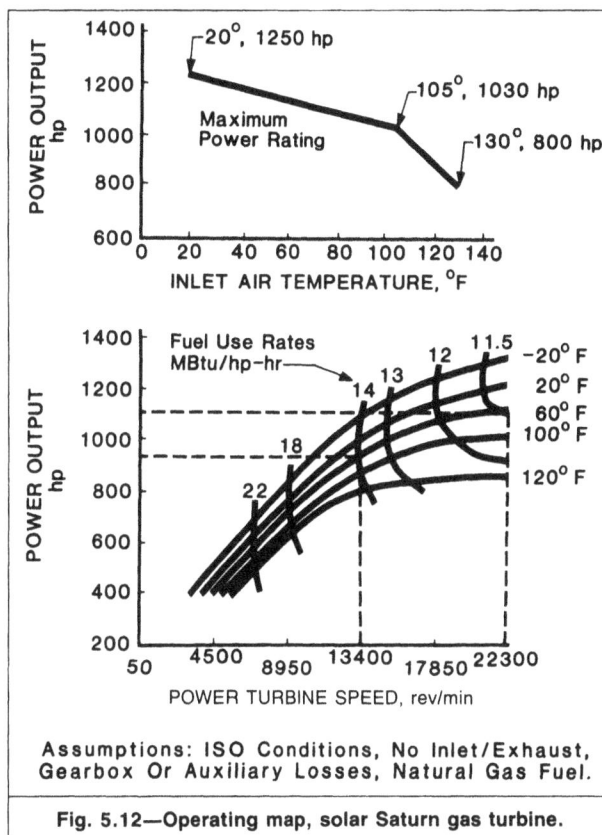

Assumptions: ISO Conditions, No Inlet/Exhaust, Gearbox Or Auxiliary Losses, Natural Gas Fuel.

Fig. 5.12—Operating map, solar Saturn gas turbine.

fuel quality and availability, some derating for wear, and naturally aspirated vs. turbocharged engines.

5.7 Single-Cylinder Units

Sec. 5.6 addressed the medium- to high-speed multicylinder engine. Another common engine used in the oil industry is usually referred to as a single-cylinder unit or "one lunger," although it sometimes has two cylinders. For many years, this type of engine was the workhorse of the industry. These are heavy, relatively slow, and long-lived units. The original investment per brake horsepower is higher, but the maintenance is lower than the medium- to high-speed engines. Until recently, the fuel consumption of about 14,000 Btu/bhp-hr was nearly twice that of the higher-speed engines. Where natural gas is inexpensive, this relatively high fuel usage is of little concern. With increasing gas prices in recent years, however, changes have been made to increase the efficiency.

An example of this type of engine is the Ajax DP-115, a single-cylinder gas engine. To smooth out the effects of a two-cycle, single-cylinder engine, the vendor installs a 3,750-lbm flywheel. Total unit weight is 13,350 lbm or more than 115 lbm/hp. Engine speed is 360 rev/min for 115 bhp. At 200 rev/min, rated horsepower is 65 hp. The unit is furnished with either a standard gas manifold or with gas injection. Marquis[10] provides fuel-consumption charts for an Ajax DP-80 that show a definite saving with gas injection. The curves in Fig. 5.10 are typical of 60-hp and larger Ajax engines but not the smaller units.

5.8 Diesel Engines

Diesel engines are uncommon in oilfield operations in the U.S. and probably throughout the world. Diesel ratings will not be covered in this monograph. The engineer should evaluate the vendors' ratings in a manner similar to that for gas-fired engines.

5.9 Gas-Fired Turbines

Gas turbines are relatively new to waterflood operations. The first commercial use is believed to have been in the Goldsmith field of west Texas in 1961.[11] Compared with an internal-combustion engine, the turbine has advantages of less weight and rotating, rather than reciprocating, moving parts. A schematic of a gas turbine is shown in Fig. 5.11. In a gas turbine, ambient air is brought to the inlet of an axial compressor that increases the pressure to a level that allows combustion. Fuel, in this case gas, is injected into the combustor, where it combines with the compressed air and is burned. The hot gases expand as they pass over two sets of turbine blades. The first set drives the air compressor and the second tur-

bine powers the driven unit. The combustion gases are still very hot when they pass out the exhaust. Heat-recovery equipment can be installed in the exhaust line to increase overall thermal efficiency. This recovered heat can be used for a number of useful purposes, such as oil treating, steam generation, and space heating.

Mechanical-drive turbine units have two shafts, with the compressor turbine operating independently from the power turbine. The power turbine is free to find its own speed as a function of the driven load. The power output of the turbine is increased by adding more fuel. The turbine will eventually settle at a power turbine speed and heat- or fuel-use rate according to the particular turbine's performance curves (map) at a given air inlet temperature. For the turbine shown in Fig. 5.12, the unit can deliver 1,100 hp at 22,300 rev/min by burning 12.9×10^6 Btu/hr of fuel.

This turbine can also provide 920 hp at 13,400 rev/min using 12.9×10^6 Btu/hr. There is a severe penalty in the latter case for operating away from the optimum speed in that it takes as much fuel to provide 920 hp as it did to produce 1,100 hp in the previous case.

A different set of curves for a larger turbine is shown in Fig. 5.13. This unit also shows the increasing fuel use per horsepower-hour at less than optimum loading. This figure demonstrates the

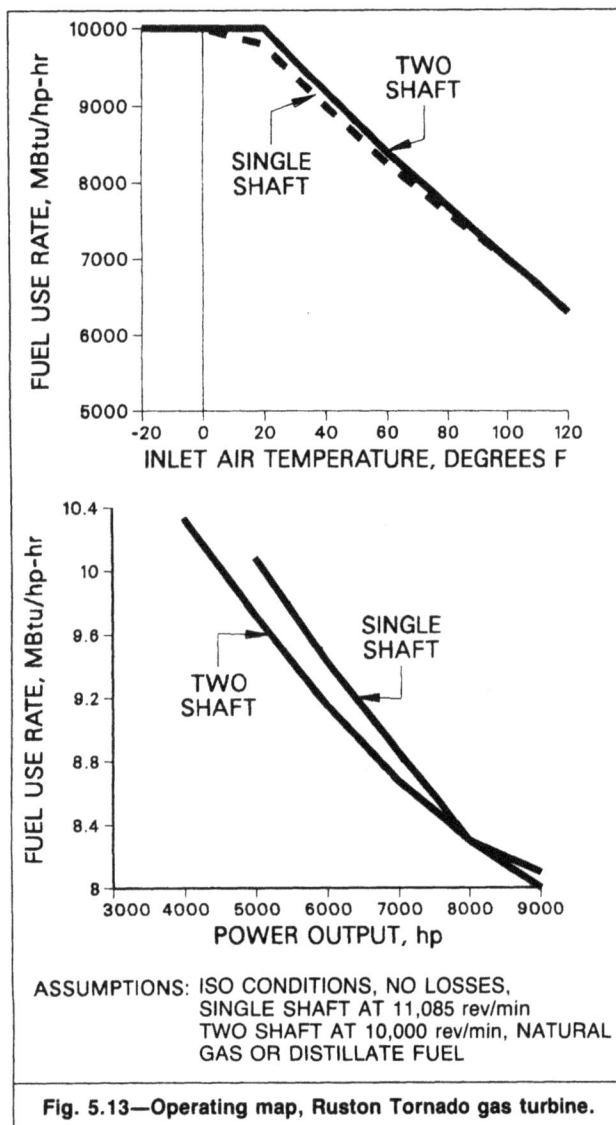

Fig. 5.13—Operating map, Ruston Tornado gas turbine.

ASSUMPTIONS: ISO CONDITIONS, NO LOSSES,
SINGLE SHAFT AT 11,085 rev/min
TWO SHAFT AT 10,000 rev/min, NATURAL
GAS OR DISTILLATE FUEL

pump operating at 3,000 to 6,000 rev/min. Some turbines operate in this speed range and can be directly connected. The high-speed gear boxes are quite efficient, with mechanical loss usually 2 to 3%.

All turbines have a characteristic that must be considered in the design. The engine accelerates from a cold start to full speed in a matter of seconds. The compressor shaft speed must approach 50 to 60% of full speed to start properly. The power shaft will accelerate at almost the same rate. This rapid acceleration causes the following events.

1. The pump is not lubricated during this short period but is already at high horsepower. The vendor should be consulted for starting procedures such as prelubricating before starting.

2. If the pump is to be started against a closed valve, an efficient control scheme is required to open the valves sequentially (see Sec. 5.13.4).

3. The pump demands water faster than the liquid can be accelerated in a static system. This problem will be discussed under suction-system piping in Chap. 6.

These problems are not unsolvable, but the solutions must be incorporated in the initial design.

The horsepower curves shown in Figs. 5.12 and 5.13 assume zero inlet and exhaust losses. Because of the high rotating speed and the huge amount of combustion air required, a turbine needs very clean air. Installation of an inlet-air plenum chamber to remove contaminants creates an inlet pressure drop. If access to cool, clean air necessitates lengthy inlet-air piping, an additional inlet pressure loss is incurred. Similarly, the hot exhaust gas must be vented from the pump room. The required exhaust piping again causes pressure drops. In most turbine systems, the inlet and exhaust pressure losses are about 4 in. of water column each. This total loss of about 8 in. of water column decreases the available horsepower. The precise amount of loss varies from one machine to another and must be obtained from the vendor for the specific inlet and exhaust system installed.

As an example of the effect of inlet and exhaust losses, the performance bulletin for a Saturn mechanical-drive turbine states that a normal installation has 2 to 6 in. of water loss on the inlet and 2 to 4 in. of water drop in the exhaust system. At near full load, available horsepower is decreased by 6 hp/in. of water from inlet losses and 3.5 hp/in. of water from outlet losses. A nominal 1,145-hp installation with a 4-in. inlet loss and 2-in. exhaust duct loss would be derated by $(4 \times 6 = 24) + (2 \times 3.5 = 7)$ or 31 hp. As with internal-combustion engines, turbines must be derated for actual conditions. For altitude, the derating factor is a direct function of on-site atmospheric pressure divided by sea-level pressure. At 3,000 ft, for example, the on-site atmospheric pressure is 13.2 psia and the derating factor is $13.2 \div 14.7 = 0.9$. The turbine must be derated by 10% of its nominal value. Combining these factors for the Saturn turbine, the unit has a nominal power rating of 1,145 hp less 10% for altitude minus 31 hp for inlet and exhaust losses for a derated power of 1,000 hp. There is a further loss from turbine-driven auxiliaries, which is normally less than 1%.

Fuel consumption is determined on the nonderated performance. If the above engine had a specific fuel consumption of 11,500 Btu/hp-hr without derating, it will have $(1,145/1,000) \times 11,500 = 13,170$ Btu/hp-hr fuel usage rate based on the actual performance of 1,000 hp.

The available turbine power is decreased or increased if the ambient air temperature is above or below the nominal temperature of 60°F. These changes affect fuel consumption per horsepower-hour in a complicated manner, as seen in Fig. 5.12.

Turbines can be adapted to use a wide variety of fuels from very low to very high-Btu gases and most liquid fuels. Experiments have even been run with pulverized coal. Turbine blades can be damaged by contaminants, including vanadium, sodium, potassium, calcium, sulfur, and lead. To prevent damage, offshore turbine installations are often routinely washed with fresh water to remove the salts. Solids in a liquid fuel are removed by filtration.

The manufacturers of turbine units can supply the specific information for their machines concerning the various factors discussed above. The engineer should use these resources to ensure the proper design. The details of the mechanical design and the relative merits of different turbines are a very complicated subject that has been covered only briefly in this section. Moreover, relatively few water-

greater efficiency of the two-shaft machine compared with the single-shaft unless the turbine is at or near full-load conditions. Figs. 5.12 and 5.13 indicate another principle of gas turbines. The full-load fuel-use rate for the Saturn at 1,120 hp is 11,500 Btu/hp-hr at 60°F compared with the Ruston at 8,500 hp using about 8,200 Btu/hp-hr. An even larger unit, the Solar-Mars, burns 7,750 Btu/hp-hr, developing 11,150 hp. Hence, the fuel consumption per horsepower-hour generally decreases with increased size and power. The engineer must consider this in choosing the turbine for a particular job. Turbines are large-horsepower drivers, rated to at least 45,000 hp in a mechanical package and much higher in the single-shaft machines.

Turbines are usually able to develop their rated power throughout their life. The power can be limited by exhaust gas temperature and by excessive deterioration of the turbine blades from solids in the air or corrosive elements in the fuel or air. Turbines have factors that make them attractive for offshore use. As rotating machines, they have excellent balance, with little vibration. A major advantage of turbines is that their weight-to-power ratio is much better than reciprocating, internal-combustion engines. The Solar Saturn weighs 8,300 lbm in a conventional mechanical package developing about 1,200 hp or 7 lbm/hp. The Ruston shown in Fig. 5.13 has a different design philosophy and weighs only 5 lbm/hp delivered. By contrast, a conventional, internal-combustion engine will weigh 15 to 20 lbm/hp.

Another factor to be evaluated is the high power-shaft speeds compared with centrifugal pumps. In some cases, a high-speed centrifugal pump can be designed for direct coupling with a power shaft. A common installation is to use a gear reducer with the centrifugal flood engineers will have expertise in this area. Hence, it is recommended that in-house or outside consultants with expertise in rotating equipment assist in the design. Turbine drivers imply a relatively large injection-pump station. The additional cost of a consultant is usually insignificant compared with the cost of such projects.

5.10 Energy Cost Considerations

Any evaluation of a prime mover must consider the present worth of the future cost of energy. Unfortunately, recent experience has shown that even experts cannot predict future energy costs successfully. Most authorities agree that the cost of natural gas will approach the world price of crude oil on an equivalent-Btu basis. In most areas, electricity cost is a function of fuel cost at the generating facility. The type of fuel used varies, with fuel oil, natural gas, coal, and even water power dominating in different areas. Understanding the high level of uncertainty involved, the engineer should consult the latest literature to estimate future fuel costs and apply these data to the local conditions.

As an example, if an engineer wished to compare a motor to an internal-combustion engine, the cost of energy for a 100-hp output prime mover can be calculated as follows. A 100-hp motor will use 81 kW at 92% efficiency. Assuming a 5¢/kW-hr electric rate, the motor costs $4.05/hr to operate. A 100-hp internal-combustion engine with a fuel rate of 14,000 Btu/hp-hr requires 1,400,000 Btu, or 1.4 Mcf of 1,000-Btu gas to operate 1 hour. At a price of $3.00/MMBtu, the engine costs $4.20/hr to operate. In this example, the energy costs for the motor and the engine appear to be about the same. The energy costs per unit shown are realistic for the 1985 time frame. Because a motor costs much less to install and to maintain, it would be the best choice in this example. This calculation assumes that the costs to provide the gas or the electricity to the site are equal. During 1982–86, gas prices declined substantially, but electricity costs generally continued to increase. If the engineer expected this type of future cost behavior, the choice of a motor might prove to be wrong, depending on the relative changes in fuel prices. Allowing for the uncertainty associated with a price prediction, the engineer must make an estimate on the basis of the best available information and evaluate the most economical choice for a prime mover.

5.11 Selection of Injection Pumps

A pump is a device used to increase the pressure of a fluid. The pressure differential between the pump discharge and a downstream point causes the fluid to move in the direction of the lower pressure. Waterflood injection pumps are almost universally one of two general types, reciprocating PD or centrifugal. The PD pumps increase pressure by trapping a fixed volume of fluid at suction pressure and then compressing that fluid to a discharge pressure. There is no flow until the discharge valve opens. Centrifugal pumps increase pressure by increasing the velocity of the fluid within the pump and converting the energy to a pressure increase at the pump discharge.

As shown in Fig. 5.14, there is a fundamental difference between these types of pumps. The PD pump has a capacity (volume of fluid pumped per unit of time) that is a function of the operational speed. For a given plunger size, the rate does not change significantly over a wide range of discharge pressure. The centrifugal pump has a capacity that is a direct function of the pressure increase across the pump. In the extreme case, the pressure can become sufficiently high that flow will cease. There is also a relationship between various pump speeds that will be discussed later.

5.11.1 General Considerations. A number of factors enter into the choice of waterflood pressure pumps, such as (1) prime mover to be used, (2) injection rates and pressures required throughout the waterflood life, (3) need for future flood expansion, (4) injection-water quality, (5) space available for pump, (6) pump efficiency, (7) initial costs, including installation, (8) operating and maintenance costs, and (9) impact of downtime.

Centrifugal pumps are often selected with gas-fired turbine drivers. The early (1961) Goldsmith project in west Texas used a turbine-driven, high-speed (6,000-rev/min) centrifugal pump,

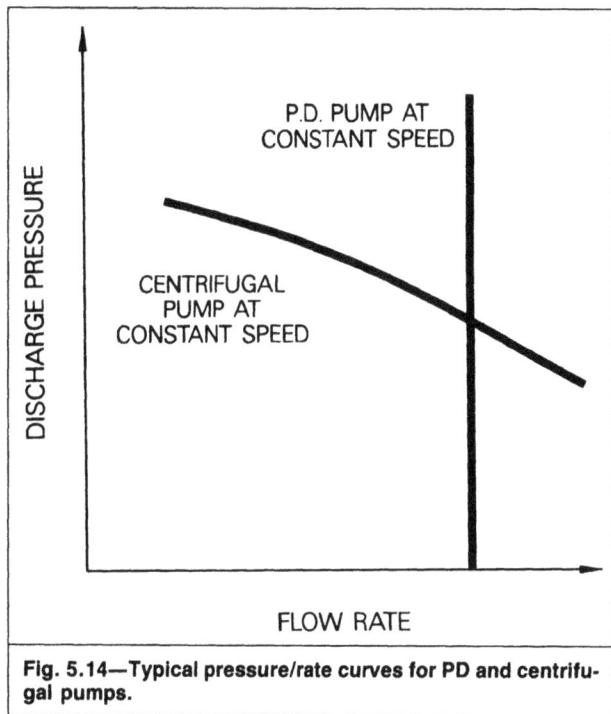

Fig. 5.14—Typical pressure/rate curves for PD and centrifugal pumps.

which supplied the desired volume and pressure more economically than a slow-speed centrifugal or a PD pump.[11] In a Cook Inlet waterflood, a centrifugal pump directly connected to a 14,000-rev/min turbine developed a high pressure of 5,500 psi operationally with a design limit of 7,600 psi.[12] Because this was a platform installation, the package size and weight were critical. In both instances, relatively inexpensive gas was available for fuel. Gas is usually initially available in a waterflood field but electricity sometimes is not. Centrifugal pumps also work well with gas engines. The variability in gear boxes coupled with the wide range of engine-speed control makes a very flexible package.

Before reservoir fill-up, water is usually injected at low pressure and maximum achievable rates. As the flood progresses, the actual injection rate may decrease, in some cases to less than half the initial value. During the same period, the injection pressure will normally increase to a level considerably above the initial pressure. When plunger pumps are used, it is common practice to select pumps that deliver large volumes of water at lower pressure with large-diameter plungers. Then the plungers are changed to smaller sizes to deliver lower volumes at higher pressures. The prime mover is sized to accommodate these changes. Because of the pump characteristic shown in Fig. 5.14, centrifugal pumps create more difficulty when the required pressure increases. This can be a serious problem, especially if the units were closely designed for the initial conditions. The possible solutions include (1) scrapping the original units and installing a new design; (2) adding stages to an originally destaged unit; (3) increasing the pump impeller diameter; (4) increasing the pump speed, if the driver can handle the additional load; and (5) operating at a higher pressure on the curve and adding new units for rate as needed.

In the early period of a waterflood, it may be possible to use the water-supply pumps to deliver water directly to the reservoir. At the Webster Field Unit,[13] two water-supply wells lifted by submersible pumps delivered about 32,000 B/D water to the injection wells through the distribution system for 9 months before completion of the water-injection plant.

Some other general factors should be considered in pump selection. Plunger pumps and some centrifugals can tolerate a certain level of turbidity. Turbidity cannot be tolerated by high-speed, high-volume, centrifugal pumps. Thus, these pumps will require the added cost of a water-treatment system (see Chap. 4). A centrifugal pump usually requires less space and weighs less than a PD pump of the same horsepower. Centrifugals normally do not need pulsation dampeners. Plunger pumps have higher mechanical efficiency than centrifugal pumps delivering the same rates and pres-

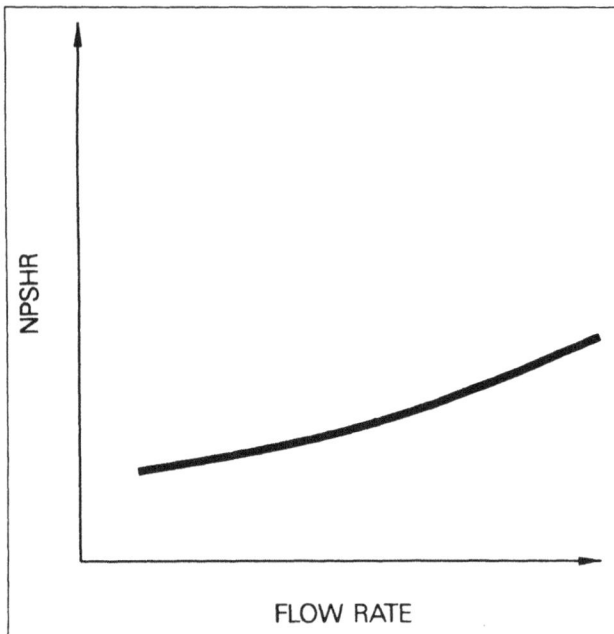

Fig. 5.15—NPSHR as pumping flow increases.

TABLE 5.5—VALUE OF CONSTANTS C_1 AND C_2

Pump Type	C_1
Simplex DA	0.001660
Duplex SA	0.001032
Duplex DA	0.000624
Quadruplex DA	0.000434
Triplex SA	0.000339
Triplex DA	0.000338
Sextuplex SA	0.000237
Quintuplex SA	0.000199
Septuplex SA	0.000146
Nonuplex SA	0.000107

Pump Type	C_2
Simplex DA	0.200
Duplex SA	0.200
Duplex DA	0.115
Quadruplex DA	0.115
Triplex SA or DA	0.066
Sextuplex SA	0.055
Quintuplex SA	0.040
Septuplex SA	0.028
Nonuplex SA	0.022

DA = double acting.
SA = single acting.

sures. Moreover, the efficiency of a centrifugal pump decreases at any condition not on its most efficient point locus. Generally, the initial investment and maintenance cost of a centrifugal pump will be less than for a plunger pump. Because of its lower efficiency, the energy costs for the centrifugal may be considerably higher.

Specifics will be discussed later in this chapter, but first it is necessary to discuss some fundamental principles that apply to both types of pumps.

5.11.2 Net Positive Suction Head. Many articles and papers have been written about net positive suction head (NPSH). As Taylor[14] said, the more written on this subject, the more confusing it seems to become. The action of any pump when drawing fluid into the suction will create a reduced pressure at that point. An external force is required to push fluid into the suction to maintain an adequate pressure. This external force on pressure is called NPSH. The effect of having inadequate NPSH available is cavitation and pressure pulses in the pump. Cavitation occurs when the suction pressure drops low enough to cause the liquid to flash to a gaseous phase. The subsequent collapse of the vapor bubble causes a brief but energetic pressure pulse in the system. These pulsations can cause problems ranging from noise to severe damage to the pump and piping. Even if cavitation does not occur, pressure pulses ("water hammer") caused by the sudden acceleration of the liquid may be present in the system, causing a gradual deterioration of the equipment.

There are two types of NPSH to be considered in pumping system design. The first is the required NPSH (NPSHR), which is a function of the pump design and flow rate. NPSHR is measured in psia for reciprocating pumps. Manufacturers will normally supply this value for their pumps. NPSHR is determined by testing to find the pressure at which cavitation starts for a given flow rate.

For centrifugal pumps, this point is determined by some test methods as the pressure at which the total dynamic head drops 3% from the maximum obtainable efficiency. Two excellent articles[15,16] on NPSH are recommended references. For PD pumps, NPSHR is similarly determined by testing. The criterion used is the value of NPSH at which the capacity is reduced by just over 3% when the pump speed and discharge pressure are held constant.[17] NPSHR increases with flow rate, as shown in Fig. 5.15. A good conservative design criterion is to add 3 to 10% to the manufacturer's maximum value of NPSHR.

The second type of NPSH is the available NPSH (NPSHA). For the pump to operate satisfactorily, NPSHA must exceed NPSHR. This type of NPSH is a function of the suction system design. Fig. 5.16 shows a typical reciprocating pump system. NPSHA, H_a, can be calculated using various parameters:

$$H_a = p_a + p_z - p_{vp} - p_f - p_{ac}, \quad \dots \dots \dots \dots \dots (5.4)$$

where

p_a = pressure exerted at water surface,
p_z = pressure head from the surface of water,
p_{vp} = vapor pressure of water at system temperature,
p_f = flowing frictional pressure losses, and
p_{ac} = acceleration pressure losses,

and where units must be consistent.

Acceleration pressure losses are a function of the action of the pump plungers. As the pump plunger starts its suction stroke, it accelerates to the midpoint of the stroke, then decelerates to come to rest at the end of the stroke. Outside force is applied to drive this action. To feed the pump suction, however, the liquid must also be accelerated. Again, an outside force is required. This force is supplied by the NPSHA and the quantity needed, p_{ac}, can be calculated on the basis of the fundamental laws of Newtonian physics. Several equations have been derived, all of which are similar except for the units. Miller[15] derived the following equation for customary units:

$$p_{ac} = (C_1 L N q \gamma)/d^2, \quad \dots \dots \dots \dots \dots \dots \dots (5.5)$$

where

C_1 = constant based on pump type found in Table 5.5,
L = length of suction piping,
N = pump speed,

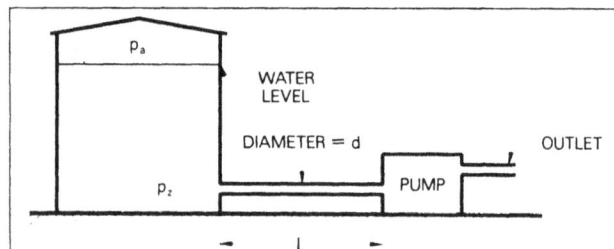

Fig. 5.16—Typical PD pumping system.

TABLE 5.6—PERFORMANCE DATA OF A336-H TRIPLEX OILWELL PUMP							
Plunger Diameter (in.)	Plunger Area (in.^{2})	Displacement (gal/rev)	Maximum Pressure (psi)	100 rev/min (B/D)	200 rev/min (B/D)	300 rev/min (B/D)	320 rev/min (B/D)
2	3.1416	0.2448	3,180	840	1,679	2,519	2,687
2⅛	3.5466	0.2764	3,180	948	1,896	2,844	3,033
2¼	3.9761	0.3098	3,145	1,063	2,125	3,188	3,401
2⅜	4.4301	0.3452	2,820	1,184	2,368	3,552	3,789
2½	4.9087	0.3825	2,546	1,312	2,624	3,936	4,198
2⅝	5.4119	0.4217	2,310	1,446	2,893	4,339	4,629
2¾	5.9396	0.4628	2,105	1,587	3,175	4,762	5,080
2⅞	6.4918	0.5059	1,926	1,735	3,470	5,205	5,552
3	7.0686	0.5508	1,768	1,889	3,778	5,668	6,046
Brake horsepower required				63	126	189	202

q = pumping rate,
γ = specific gravity of liquid, and
d = ID of suction piping.

Hence, acceleration pressure is a direct function of pump speed, pumping rate, and suction-pipe length but an inverse-squared function of suction-pipe diameter. This relationship demonstrates why a pump may fail to perform properly when the flow rate is increased and/or the pump speed is increased. This equation also demonstrates the need for a good suction-piping design. Table 5.5 shows that acceleration pressure is inversely proportional to the number of plungers used in the pump. Evidence indicates that Eq. 5.5 will yield conservative (high) values, particularly as suction-line length increases.

Another equation for acceleration head was developed empirically[18] as:

$$p_{ac} = (LvNC_2)/(Kg), \dots\dots\dots\dots\dots\dots\dots\dots (5.6)$$

where

v = velocity of fluid,
C_2 = constant based on pump type (see Table 5.5),
K = constant based on fluid type, 1.4 for water, 2.5 for hot oil, and
g = acceleration of gravity.

As an example, if a triplex pump is running at 300 rev/min pumping 34,000 B/D of fresh water with a 10-in.-ID suction line that is 50 ft long, then the acceleration head can be calculated as:

$$v = \frac{(34,000 \times 5.615)/(1,440 \times 60)}{(5/12)^2 \times 3.14159} = 4.05 \text{ ft/sec}$$

and

$$p_{ac} = \frac{50 \times 4.05 \times 300 \times 0.066}{1.4 \times 32.17}$$

= 89 ft or 38.5 psi.

This acceleration head value is a comparatively large number because p_f for this system would be on the order of a few tenths of a psi. The acceleration pressure could be reduced by changing the suction-piping size or length, but the effectiveness of this approach is limited. For example, if the piping above were to be 20 in. in diameter, then p_{ac} would equal 22 ft or 9.6 psi, which is still a large number. The most effective means to reduce acceleration pressure to an insignificant value is to install a pulsation dampener at the pump suction. Miller[19] discusses the types of dampeners, effectiveness of the dampeners at different frequencies, and sizing considerations and is recommended to the waterflood engineer. The pulsation dampener that appears to be most effective for low frequencies (<50 cycles/sec) is the energy-absorbing, gas-bag, flow-through type. Several versions of this type are commercially avail-

able. The combination of proper suction-piping design (short, large diameter) with a correctly designed pulsation dampener will produce a low acceleration pressure with little piping vibration.

Going back to Eq. 5.4, NPSHA can be calculated for any given system design. However, determining whether the value is adequate requires knowledge of NPSHR. Unfortunately, this parameter is not known until the pump has been selected. The waterflood engineer frequently needs to estimate the NPSHR at an earlier stage to work on preliminary piping and facility design. Langewis and Gleeson[20] suggest estimating NPSHR (in feet) as equal to the liquid vapor pressure head plus friction losses plus acceleration head plus the elevation (in thousands of feet). The elevation factor is to allow for the decrease in atmospheric pressure that is necessary because NPSH is in terms of absolute pressure. NPSH estimated by this method will tend to be conservative, but the engineer can use this value as a starting point for the design. Again, in calculating NPSHA, the pressure acting on the surface of the liquid in the suction tank must be included. This is frequently missed initially and can lead to designs that require the suction tank to be elevated or pumps to be installed below ground level. The engineer is cautioned to calculate NPSH in terms of absolute pressure.

As stated previously, NPSHR is normally determined by the centrifugal-pump manufacturer as the value of NPSH at which the total dynamic head output decreases by 3% at rated speed and flow rate. Taylor[14] points out that this value is not necessarily the point at which cavitation begins. In fact, cavitation may or may not be occurring at this point, but it is assumed to be insufficiently severe to have a serious effect on pump life. Taylor goes on to show that high-energy (650-ft/stage) pumps at low flow rates can suffer damage. Hence, such pumps should not be substantially oversized or run at low rates. The suggested safety factor of adding 3 to 10% to the manufacturer's NPSHR may not always be practical. The engineer must evaluate the cost of the damage caused by cavitation compared with the cost of getting a higher value of NPSHA so that the correct economic decision is made.

5.12 PD Pumps

PD pumps are the workhorse pumps for waterflood operations. Historically, PD pumps have been ruggedly built to stand up to the use (and abuse) of a waterflood installation. A PD pump usually is the "best" candidate for low-volume, high-pressure floods. However, the evaluator must be aware of the limitations of this type of pump. The injection rate is a function of pump speed and plunger size. Although both can be changed to cover a broad range, any specific plunger size has a definite pressure limitation. Table 5.6 is a typical presentation of pump specifications. It can be seen from these data that changing from 2⅜- to 2½-in. plungers will increase rates from 3,552 to 3,936 B/D water at a constant pump speed of 300 rev/min. But, if the system requires a 2,750-psi discharge pressure, the larger plungers cannot be used. Hence, while rates can be varied, pressures may limit the range. The engineer must evaluate the effect of both injection pressure and rate on pump design.

Manufacturers' rates for pumps are usually shown at 100% volumetric efficiency, as in Table 5.6, but may be stated at 95% or some other efficiency. The engineer must ascertain what basis is

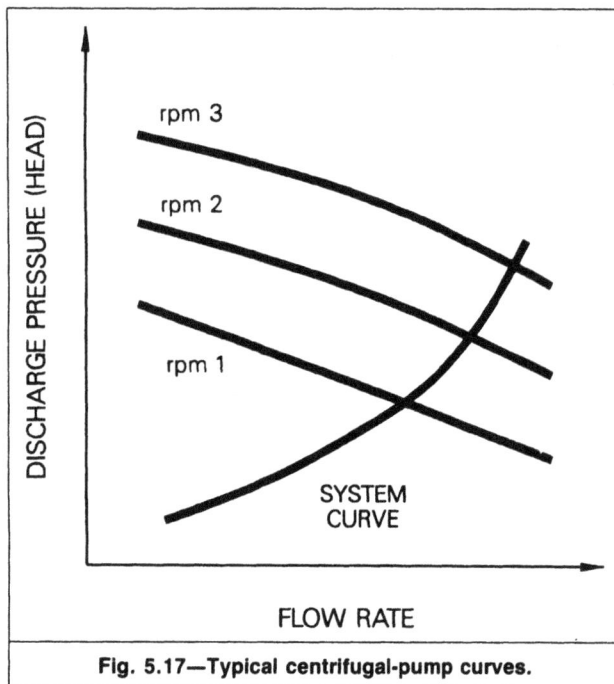

Fig. 5.17—Typical centrifugal-pump curves.

being used. A 90% volumetric efficiency is a reasonable value to use to allow for wear. So, from Table 5.6, the 2⅜-in. plunger running at 300 rev/min can be designed for 3,200 B/D water and the 2½-in. for 3,500 B/D water. Similarly, there is a mechanical efficiency loss that must be accounted for, even though this is frequently not explained in vendors' literature. Generally, a 5% mechanical loss will occur with either a belt drive or a gear box to couple the pump and prime mover. Again from Table 5.6, 189 bhp required at 300 rev/min would actually need a 200-bhp driver. Note that this size driver would not be adequate at 320 rev/min.

The manufacturers of PD pumps commonly use a nomenclature similar to the following example (as used by the Oilwell Div. of USX Corp.): the first digit in the model number shows how many plungers, the second is the maximum plunger diameter in inches, and the third is the stroke length in inches. Thus, a Model 334 is a triplex with a maximum of 3-in.-diameter plungers and a 4-in. stroke. Unfortunately, this model number gives no indication of the rated horsepower.

PD pumps are built with an odd number of plungers. Common sizes are triplex and quintuplex pumps, with three and five plungers, respectively. Rarely, pumps have been built with up to nine plungers, a nonuplex. The plungers travel through replaceable cylinder liners. The metallurgy and finish of the liners are as important as they are for the plungers. Any change in plunger size may require that the liners also be replaced. This sizing should be considered as part of the original design. Generally, waterflood PD pumps are single-acting, meaning that there is only one pressure stroke per plunger during a full (360°) rotation of the crank shaft. The pumps may be built with the plunger travel in the horizontal or vertical plane. There are several major differences between horizontal and vertical pumps. The plunger of a horizontal unit is pushed by a connecting rod on the power or pressure stroke. The plunger of a vertical unit is usually pulled down on the power stroke by a pair of rods extending beyond the plunger. This arrangement requires somewhat more attention to alignment. As might be expected, the vertical unit takes up less floor space and has the advantage that the major forces acting on its base are vertical. The horizontal unit can impose high horizontal loads on its base. Because the suction and discharge flanges of the vertical unit are at the top of the pump, there is additional unsupported piping compared with the horizontal type. The latter is also usually easier to work on during repairs or maintenance operations.

5.12.1 Advantages/Disadvantages. The advantages of a PD pump compared with centrifugal pumps are (1) a high volumetric and mechanical efficiency; (2) lower fuel costs resulting from higher effi-

ciency; (3) increased power demand as a direct function of speed; (4) greater tolerance to adverse water quality owing to a wide choice of plunger and liner materials; and (5) the rate is not affected by pressure over a broad range. These must be weighed against the disadvantages: (1) the unit will usually have higher initial investment and maintenance costs; (2) a plunger pump creates potentially destructive pulsations that might require dampening (at additional cost); and (3) a higher throughput requirement, necessitating a larger-diameter plunger, may not be achievable because of pressure limitations.

The investment for PD pumps is greater because of the higher unit weight and size. As a rule of thumb, the maintenance cost for a PD pump is about three times that of a centrifugal pump on a per-unit power basis.

Pulsation in PD pumps results from the open/close sequence of the valve, which causes flow acceleration and deceleration in a sinusoidal pattern. For a single-acting duplex pump, maximum fluid velocity in the discharge piping is about 160% of the average value. This variation decreases to a range of 82 to 107% of average velocity in a triplex and only 94 to 102% in a quintuplex. The pulsations in the flow of PD pumps will tend to synchronize when several pumps are installed in parallel. This will be especially evident on electric-motor-driven units at similar speeds because the motor slip will adjust to the load. If the PD pumps do synchronize, the effect of the pulsation will become additive. Potentially catastrophic vibration and pressure surges can occur. Multiple pulsation dampeners or wide separation of the pumps are recommended in such cases.[19]

Henshaw[17] states that most problems with a PD pump can be avoided by selecting a pump that will operate at conservative speeds, carefully designing the piping system, and using maintenance practices that preserve the alignment of the plunger and stuffing box. This alignment is important because packing problems can be significant. As with most equipment, operation at maximum speeds or loads will usually increase maintenance costs drastically.

5.13 Centrifugal Pumps

It is probable that all inexperienced engineers incorrectly specify centrifugal pumps. Perhaps the terminology is confusing, or perhaps the concepts are. This section is intended to alleviate this problem. If the engineer continues to have difficulties, expert personnel and pump manufacturers should be consulted.

5.13.1 Advantages/Disadvantages. Centrifugal pumps have advantages and disadvantages compared with reciprocating units. The greatest disadvantage of centrifugal pumps is their low efficiency compared with PD pumps. Centrifugal pumps are not designed to be effective at low rates and high discharge heads.

A centrifugal pump increases pressure by greatly increasing the velocity of the liquid as it passes through the impeller and then converting that velocity head to pressure. Where a PD pump generates the discharge pressure in one stage, a centrifugal can be multistaged. For example, a pump capable of 3,500 to 4,000 psi discharge pressure may have 10 stages.

Because the centrifugal creates velocity that is converted to pressure, the discharge pressure may be increased by increasing pump speed. As discussed earlier, doubling the speed would quadruple the pressure because the discharge pressure is a function of the rotating speed squared. However, this relationship makes the power requirement a cubic function of the speed. These facts can be expressed as

$$q \propto N, \dotfill (5.7a)$$

$$p_d \propto N^2, \dotfill (5.7b)$$

and

$$P_{hp} \propto N^3, \dotfill (5.7c)$$

where P_{hp} is the input power required.

Increasing a centrifugal pump speed increases the rate if the pressure is held constant. If the rate is held constant, a speed increase will cause the pressure to increase as a squared function. In both

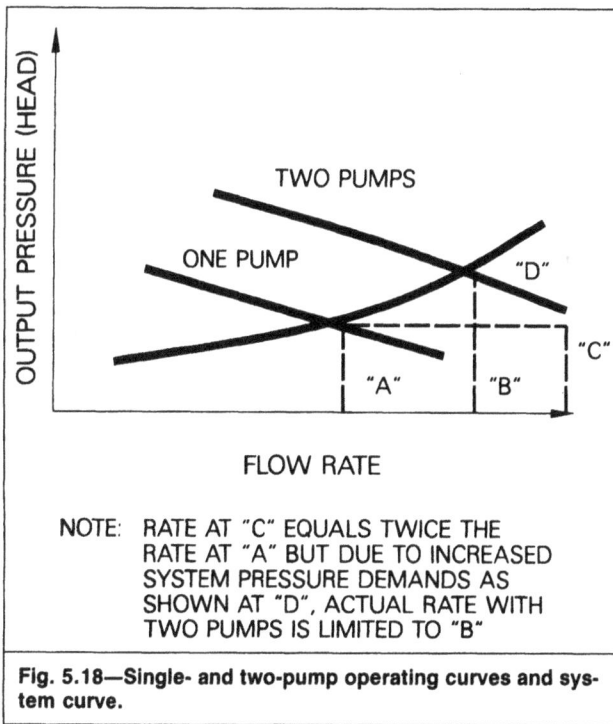

Fig. 5.18—Single- and two-pump operating curves and system curve.

NOTE: RATE AT "C" EQUALS TWICE THE RATE AT "A" BUT DUE TO INCREASED SYSTEM PRESSURE DEMANDS AS SHOWN AT "D", ACTUAL RATE WITH TWO PUMPS IS LIMITED TO "B"

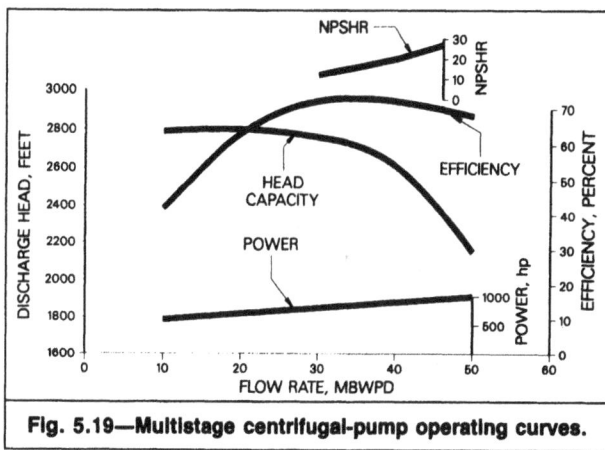

Fig. 5.19—Multistage centrifugal-pump operating curves.

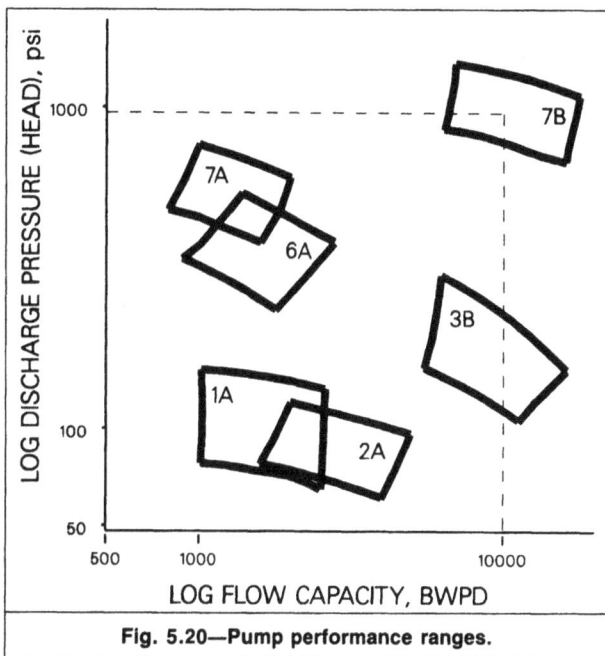

Fig. 5.20—Pump performance ranges.

cases, power increases as the cube of the speed ratio. These relationships are shown in their simple forms in Fig. 5.17, which shows that both rate and head can be changed simultaneously.

A centrifugal pump will operate at the point where the head curve intersects the system curve. This may not be the intended condition. Therefore, the design must anticipate this point. The system curve is the result of all pressure requirements downstream of the pump discharge flange, including valves, fittings, pipe, elevation changes, and wellhead pressure. The head-rate curve is a function of the difference in pressure between the suction and the discharge flanges at various flow rates. Karassik[21] maintains that a velocity term should be added to the suction side. In normal waterflood installations, this value is insignificant and can be ignored. Fig. 5.17 shows that as pump speeds are varied, the intersection with the system curve changes. The system curve can be moved upward to decrease flow at reduced speed, but generally cannot be moved downward. A centrifugal pump should never be operated to the right of the vendor's published curve without the manufacturer's concurrence. The pump is not designed for that area, so the curves cannot be extrapolated.

The head-rate curves for pumps operating in parallel are additive as to rate. If one pump handles 10,000 B/D water at 1,000 psi, then two pumps of that type are capable of 20,000 B/D water at 1,000 psi, if there is no change in the system curve. In fact, though, the system curve at 20,000 B/D water will be considerably higher than at 10,000 B/D water. Thus, if the system curve showed 1,000 psi at 10,000 B/D water, it might require 1,500 psi at 20,000 B/D water. The pumps will not be capable of the doubled rate at that pressure, as shown schematically in Fig. 5.18.

At low volumes, centrifugal pumps have low efficiency. This is shown on the performance curve for a six-stage centrifugal pump (Fig. 5.19). This particular pump shows 72% efficiency at just over 1,100 psi and about 40,000 B/D water. But the efficiency drops to 63% at 20,000 B/D water and 1,200 psi. As a comparison, most plunger pumps are rated at 92 to 97% efficiency over their entire range of performance.

The initial cost of the centrifugal pump may be half the cost of a plunger pump with the same capacity. If special corrosion trim is needed, the cost of the centrifugal may increase by up to 50%. Pump foundations usually weigh about two to three times the weight of the pump being supported. The much greater weight of the PD pump will cause its installation costs to be much higher than for a comparable centrifugal pump.

Summarizing, the centrifugal has the following differences from the PD pump.

1. Flow rate is a function of system head. This can be a disadvantage but does allow for flow-rate control without changing pumps.

2. Flow is usually relatively smooth with low pulsation.

3. The pump is mechanically simple and easy to maintain.

4. Installation and maintenance costs are low.

5. As a rotating unit, it can be sized for and directly connected to a driver, especially with an electric motor.

5.13.2 Centrifugal-Pump Selection. There are many different types of centrifugal pumps, and there may not be consistency among manufacturer's terms. Generally, the engineer will have several choices to make, including (1) single (one impeller) or multiple stage, (2) single (one suction flange) or double suction, (3) horizontal or vertical split case (normally only for one or two impellers), and (4) barrel case for pressures above 4,000 psi.

Additionally, for any given type, several options are usually available, such as volute or diffuser exits from the impellers; various metallurgies; and open, semi-enclosed, or enclosed impeller designs. The manufacturers publish type curves, such as shown in Fig. 5.20, that display their particular pump envelopes over a log/log rate-head grid. This curve shows that 10,000 B/D water at 1,000 psi total head could be satisfied by Pump 7-B in this particular pump family. Even though this might be the final pump, the engineer should request that the vendor recommend the best pump and, if desired, possible alternatives. A good practice is for the engineer to specify the conditions and desired performance and let the vendors specify the pump. The quotation can then be compared with the engineer's previous analysis for evaluation.

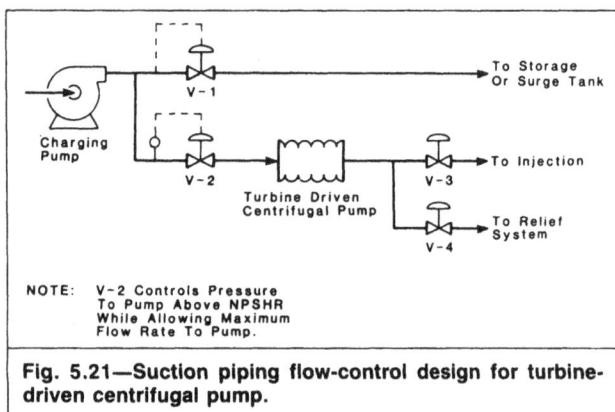

Fig. 5.21—Suction piping flow-control design for turbine-driven centrifugal pump.

In Fig. 5.20, the vertical envelope for each pump covers a range of impeller sizes, with the uppermost curve in the pump envelope representing the largest impeller and the lowermost curve representing the smallest impeller recommended by the manufacturer for that particular case. Thus, a centrifugal-pump case might accept a 9-in. impeller as the largest and a 7-in. impeller as the smallest acceptable for good performance. This represents an alternative to speed control. A pump can be purchased with a smaller impeller installed. When pressures increase, the original impellers could then be replaced with larger-diameter impellers. The engineer should remember that power requirements will increase with this change and the driver must be able to supply it. Again, power will vary as the cube of the diameter ratio, head as the square of the diameter ratio, and rate directly as the diameter ratio. These relationships can be expressed as "affinity laws":

$$q \propto d_{pi}, \dots\dots\dots\dots\dots\dots\dots\dots\dots\dots\dots\dots\dots (5.8a)$$

$$p_d \propto d_{pi}^2, \dots\dots\dots\dots\dots\dots\dots\dots\dots\dots\dots\dots (5.8b)$$

and

$$P_{hp} \propto d_{pi}^3, \dots\dots\dots\dots\dots\dots\dots\dots\dots\dots\dots\dots (5.8c)$$

where d_{pi} is the pump impeller diameter.

The standard pump line from most vendors will not require any verification as to internal design strengths. The engineer should ensure that the inlet and outlet flanges or connections match the piping design and the company's specifications. Vendors commonly supply flat- or raised-face flanges as standard, but the piping design may require ring-joint flanges. The engineer should specify the pipe connections needed for each pump.

As general guidelines, standard iron, steel, or bronze internals can be used for noncorrosive waters. For mildly corrosive water, bronze-fitted pumps should be used. Aluminum bronze, Ni-resist, or stainless-steel internals are appropriate for use in corrosive or severely corrosive waters. The Natl. Assn. of Corrosion Engineers' Recommended Practice RP-04-75 is a useful reference for the selection of pump metallurgy.[22] The engineer should supply the pump vendors with the necessary information (see below) and request their recommendations for metallurgy. The recommended material can then be discussed and final decisions made when the pump is ordered. For particularly severe conditions, the engineer should consult with metallurgical specialists.

When specifying the pump requirements, the engineer should fill out a data sheet, either from a company standard or one supplied by a vendor. At a minimum, the data listed below should be included.

1. The calculation for the discharge head required.
2. The calculation for NPSHA.
3. The type of driver preferred or required and driver output speed with as much detailed information as is known.
4. Description of the water or a water analysis (if available), including water temperature and dissolved gas content.
5. Any special features or materials required.
6. Whether a performance test is required.

7. A number of copies of drawings and manuals.
8. Spare parts required.

5.13.3 Vertical Centrifugals.
Although most centrifugals have shafts in the horizontal plane, some are vertical. Among the pumps in this classification are vertical turbine pumps (not to be confused with a gas-turbine-driven centrifugal), canned pumps, and electric submersible pumps. Although the performance curves for these pumps may differ somewhat from a "standard," or horizontal, centrifugal design, the same general principles apply. Holland[23] states some of the advantages and disadvantages of these pumps: "Vertical turbine pumps should be given particular consideration where design requirements are in the middle range of pressures and volumes. They can be well adapted in pilot waterfloods or in floods with anticipated short life. Some of the advantages of the vertical turbine pump are low first cost, low installation cost, small space requirements and generally low maintenance, at least over a short span of life. However, it is possible that maintenance over a long period of time may result in higher costs and excessive downtime. A particular disadvantage of this type of pump is its relatively low efficiency."

Johnson and Vandevier[24] discuss the use of an electric submersible well-production pump in injection service. The data indicate that a standard submersible pump driven by a standard NEMA-type electric motor in a horizontal surface installation has some economic advantage over normal PD pump installations. The high power costs associated with low efficiency are more than offset by the reduction in maintenance and repair costs. Johnson and Vandevier's paper gives a good description of the advantages and disadvantages of different pump systems and is recommended for the waterflood engineer's reference.

Another special centrifugal pump is for low-volume, high-head requirements. The pump is a single- or double-stage, very high-speed (up to 25,000 rev/min), straight-vaned impeller design. The system can pump from 3,000 BWPD at 700 psi to 10,000 BWPD at 4,500 psi. The single-impeller design in this centrifugal pump generally causes the peak efficiency to be 75% or lower. This may be offset by the simple installation, low weight, and lack of vibration compared with a PD pump for this service.

5.13.4 Suction Piping for Centrifugal Pumps.
Once a centrifugal pump has come up to speed, flow through the suction piping is constant. There is, therefore, no acceleration head present after startup of the pump. Pulsation dampeners are rarely needed. Friction head is based on rate, and NPSHA is easily determined as in Sec. 5.11.2. However, there is a special problem with starting turbine-driven centrifugals. The water in a long suction line cannot be accelerated as fast as the turbine and pump accelerate. This can be overcome by using a piping scheme like the one shown in Fig. 5.21. This scheme requires that water flow past the turbine-driven pump suction before the turbine is started. The high-speed pump usually requires a charge pump to meet NPSHR. Starting against a closed valve will also prevent the pump from throwing water out of the impeller eye.

In practice, the charge pump is started and pumps through the first valve, V-1. After a time delay, if a satisfactory pressure exists upstream of V-2, then that valve is opened and the turbine is started. V-3 will open on discharge pressure, and V-1 will close because the pump has reduced the line pressure. Because the water is passing through V-1 at startup, only the water in the short suction line at V-2 needs to be accelerated. If the distance from the charge pump to the turbine pump is long and the bypass line is not installed, the water in the piping cannot accelerate in time and the turbine-driven pump will shut down on low suction pressure.

5.13.5 Pump Seals.
Centrifugal pumps with mechanical seals experience little or no leakage. Plunger pumps often have some leakage problems, particularly as they wear. The finish of the wear surfaces of the plungers and packing glands is an important factor in the life of the seals. Company standards should be consulted for the quality of the finish that is required for these items. One major company has a standard (roughness average) of 16 μin. for plunger finish and 63 μin. for packing-gland finish. The engineer should

consult the current information available and discuss the subject with the vendors. The downtime that is required to maintain the pump in satisfactory operation is a factor that must be considered in economic analysis. Centrifugal pumps generally show less downtime. Some waterflood operators maintain a standby plunger pump for each three or four pumps in use. This is an added cost for the PD pump systems that might be justified depending on the amount of downtime expected and how critical the maintenance of constant injection rate is considered to be. An inventory of parts can minimize downtime and should be considered, especially for long-delivery items. This is particularly true when the pump supplier does not maintain a supply of such parts at a convenient location.

5.14 Injection-Station Design

A complete discussion of an injection-station design is beyond the scope of this monograph because of the large number of possibilities. Therefore, only the highlights will be covered. These include pump and driver selection, piping design, number and location of stations, controls, tanks, and other associated facilities. As in all steps of waterflood-system design, the engineer must keep in mind the importance of doing things right the first time because a bad design can ruin the economics of the project.

5.14.1 Pump Selection.
Secs. 5.12 and 5.13 outlined some general principles for PD and centrifugal injection pumps. Once the injection rate and pump discharge pressures are calculated, the first task is to select a pump type. In the 1950's and earlier, most floods used small (up to 125 hp) PD pumps. Since then, design has tended toward use of a split-case, multistage, centrifugal pump in larger-volume applications. These can produce pressures up to 4,000 psi when operating at 6,000 rev/min. As stated earlier, centrifugals usually are not competitive in high-pressure, low-rate applications. PD pumps with small plungers are commonly used for this service.

A centrifugal is capable of a wide range of injection rates without changing the equipment. A PD pump, particularly when driven by an electric motor, has little flexibility without recycling water and, consequently, wasting energy. Whether a PD or a centrifugal is chosen, future expansion must be considered. One question that needs to be answered is whether the pump will be the only one needed during the life of the flood. If it is, the pump should be able to adapt to expected changes. If it's not, plans must be made for the original pump(s) to be removed from service and replaced when conditions change. The answer may indicate, for example, that a 200-hp pump should be installed even when a 100-hp unit would be sufficient for startup.

When selecting a pump, the engineer must also consider the auxiliary devices needed. Will a charge pump be needed to supply NPSHR? Are pulsation dampeners required? (Suction and discharge dampeners are usually good investments with any PD pump.) Will a prelube system be needed with a high-speed centrifugal? What control and control sequencing will be needed to start the pump?

Some mechanical considerations must also be taken into account. Is special metallurgy needed to handle the water because of solids content or corrosivity? What is the maximum pressure rating needed? For a PD pump, at what percent of the vendor's maximum speed should the pump be run (preferably <100%)? What derating factors should be applied to the vendor's volume or pressure values? What is the pump efficiency? What will happen to efficiency or other factors with wear or change of conditions? Is the unit a prototype or relatively untested design that could have unknown defects? (It's generally a good idea to stay away from these.)

During the selection process, it must also be determined whether the pump will be treated separately, with the driver to be selected later, although tacitly taken into account during pump selection. Alternatively, will the pump and driver, including the coupling device, such as belts or gear box, be purchased as a package? The answer to this will decide how to write specifications for quotations and further discussions.

Another important consideration is whether different areas of the field will have different injection requirements. The reservoir study may show, for example, that the north end of a reservoir requires 2,500 psi pressure while the south end needs only 1,000 psi for

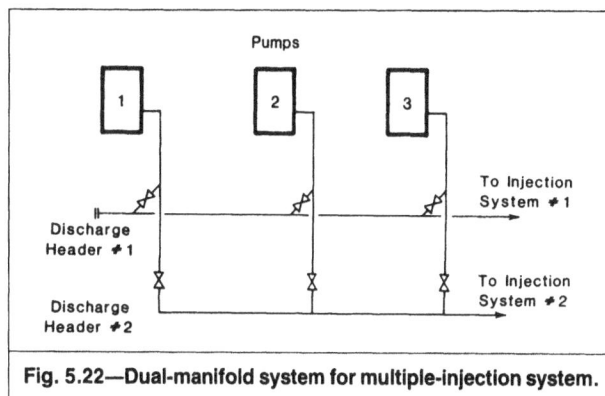

Fig. 5.22—Dual-manifold system for multiple-injection system.

adequate injection rates. Considerable horsepower can be saved by designing two pumping systems in such a case. A simple way is to connect multiple pumps to a two-system header (see Fig. 5.22). As volumes and pressures change, the needed adjustments can be made to the pumps without revising the piping.

Vendor qualifications also need to be checked. Is the vendor on an internally approved company list? What is their reputation? Is the vendor financially sound? Is the local representative known to be honest, technically competent to help in pump selection, and willing to refer the problems to the manufacturer? Where are pump service and spare parts available? How long will delivery of parts or service personnel take? Is service known to be high quality? Does the manufacturer have a reputation for delivering on time, at the quoted cost?

5.14.2 Driver Selection.
Secs. 5.5 through 5.9 outlined general principles in driver selection, including electric motors, multicylinder internal-combustion gas engines, single-cylinder units, diesel, and gas-fired turbines. The final choice will depend largely on the pump selection. Because of this dependency, it is frequently best to use the same vendor for the pump and driver. As outlined earlier, some considerations in the selection process are how much power is required; what energy sources are available, in what quantity, and at what cost; what types of equipment are field personnel familiar with and which spare parts are stocked; and how much downtime can be expected under field conditions and is a spare unit advisable.

The obvious first step is to determine how much power is needed for the whole project. This can be estimated by Eq. 5.3. This total power requirement, in combination with pump selection, can be divided into reasonable blocks. As an example, if the total power requirement is about 900 hp and 200-hp pumps are envisioned, then five units are needed. This size of prime mover is available in standard gas engines or near-standard electric motors. But the engineer must decide whether three 300-hp or two 500-hp or even one 1,000-hp unit(s) would be a better buy. The 300-hp units could still be standard engines or standard (but special-order) motors. The 500-hp units could be engines or specially built motors. The 1,000-hp size is in the range for consideration of a small gas turbine or a standard engine.

There is no hard and fast rule on how many units should be installed for a given total power requirement. For any particular type of prime mover, cost per unit power decreases with size. Some engineers prefer to have at least two 50% units to be reasonably sure that the flood is never entirely down. This is most critical when operating in cold climates because of freezing problems or when injecting produced water, which may have no other means of disposal. Other engineers prefer two 67% or even 100% units, which, of course, increases the cost. Two similar units of the same configuration should have about the same total downtime as a single unit in terms of injected barrels. As a rough estimate, for example, gas engines on PD pumps will experience about 5% downtime for each pump and driver. So, two half-capacity units will lose the same number of injection barrels as one full-capacity unit. As stated above, if two or more pressures are needed in the system, multiple units will be required. Except for a small waterflood, the largest-horsepower units that are suitable generally will be the most

TABLE 5.7—COSTS TO CONSIDER WHEN COMPARING ECONOMICS OF GAS ENGINES AND ELECTRIC MOTORS IN WATERFLOOD OPERATIONS[25]

Costs of gas engines and electric motors
Initial installation
Safety controls
Starting equipment
Maintenance costs (dependability and availability)
Operating costs (energy and labor)
Fuel lines
Cooling-water supply
Electric lines
Transformer banks
Special equipment for engines or motors
Operating costs for auxiliaries
Ad valorem taxes and insurance

economical choice, with serious consideration given to using at least two units.

All drivers have auxiliary equipment. Some important points to be evaluated are that (1) the cost of the motor control may equal or exceed the purchase cost of the motor itself; (2) most gas turbines require fuel gas at pressures in excess of 200 psi, which usually means that fuel gas compressors must be added to the system; and (3) the pump and driver must have some type of foundation, which will vary in size and cost depending on the type of system. The gas engine and horizontal PD pump will require a massive foundation block to protect the system, whereas a gas-turbine centrifugal pump can be set on a slab.

When both gas and electricity are available, an economic analysis should be made to compare the costs of each. Stevenson[25] presented a discounted cash-flow method applied to gas engines vs. electric motors. Table 5.7 is a check list for cost items to consider when making an economic study of this type. In the example given in Stevenson's analysis, there is very little difference in cost between using gas engines and using electric motors. There are a number of factors to consider that may not be easily quantified for use in the economic analysis. Casinghead gas from the formation being waterflooded is not a reliable source of gas because the supply will diminish as the flood progresses. During the flood life, the water-injection rate and the pressure requirements may vary from year to year. Future value of the gas may be much greater than the current price would indicate. In the past, electric utility companies have entered into contracts that decreased the unit cost of electricity at higher rates of consumption. In some areas, the utility companies have now eliminated this discount. The engineer must evaluate current and expected future costs and supply for both electricity and fuel gas. If a high level of uncertainty exists, it may be advisable to allow for changing the energy source at some time in the future.

The initial cost of gas engines is about two to three times higher than electric motors. So, when gas engines are selected, the initial cash outlay is considerably greater than for motors. This may be an important consideration. For waterfloods of short duration, the effect of initial investment cost on the overall economics is greater than for longer-term operations. In other cases, the economics favor gas engines, as in the Webster Field Unit.[13]

A number of waterflood operations have switched from gas to electricity as gas prices have increased. For example, at Shell's Ventura flood,[26] the WP-4 plant with a design capacity of 60,000 B/D water at 5,000 psi uses 4,500-hp centrifugal pumps with electric motor drivers. This plant was placed in operation in Jan. 1973. The other three plants, which were installed earlier, use gas as fuel. Careful engineering analysis of the particular situation at hand will determine the most economical type of prime mover for waterflooding. This analysis must give realistic consideration to future requirements and energy prices during the entire life of the waterflood.

5.15 Suction Piping

Injection-piping design will be covered in more detail in Chap. 6. Suction-piping design recommendations are listed in this section. For good results, the following suction design principles should be used. The same basic principles apply to injection-piping designs.

1. The suction system should be as simple and streamlined as possible.

2. The pump should be located as close to the surge or suction tank as possible.

3. Each pump should have its own surge tank or at least its own suction line from the tank.

4. Pipe fittings should be kept at a minimum. No abrupt high-angle changes in the direction of flow should be allowed. "Ells" should be at either 45° or long-radius 90°. In many cases, specially fabricated sweeps are most useful for changing flow direction.

5. Full-opening valves should be used.

6. The pipe from tank to pump should slope with an even gradient. Avoid piping designs that allow pockets where air or gas may accumulate.

7. The suction-line outlet at the tank should be higher than the fluid end of the pump. If this is impractical, the suction pipe should be level or should slope up to the fluid end of the pump and be connected with an eccentric reducer, belly down.

8. For streamlining, the suction-line outlet at the tank should be rounded and no pipe should protrude into the tank.

9. The suction line should be several sizes larger in diameter than the pump inlet and should be sized so that the water velocity is 3 ft/sec or less for centrifugal pumps and 2 ft/sec or less for reciprocating pumps. Flow velocity is calculated as

$$v = 0.408q/d^2 \dots \dots \dots (5.9)$$

for gal/min and inches; 0.408 becomes 212.2 for m^3/min and cm.

Although adherence to these principles is most critical when dealing with high-pressure systems using reciprocating pumps, they do apply to any waterflood plant design. The flow path from the tank to the pump suction should be as smooth and undisturbed as possible.

5.16 Location and Number of Injection Stations

Ideally, a waterflood injection station should be centrally located among the wells to which it delivers water. The obvious reason for this is the desirability of having the pump station at the shortest average distance from water-injection wells. This will minimize the total pipe length required to serve the wells. Shorter length is desirable to reduce line costs and to minimize the friction pressure loss between the plant and wells. In an actual waterflood, this principle will be altered to avoid natural hazards such as possible surface flooding, landslides, and special topographic features. Location selection should also consider the water-supply locations, roads, electrical installations, and production facilities. This frequently results in the station being sited near a centrally located production battery.

The number of injection stations may vary from one for each injection well to one central station for all the injection wells in even a large property or unit. An example of the former is the LL-5 Flank waterflood at Lake Maracaibo,[27] where each of eight injection wells has its own injection station, complete with supply pump, deaeration tower, centrifugal pump, and controls capable of delivering 60,000 B/D water. The latter case was illustrated at the large SACROC Unit in Scurry County, TX.[28] This unit encompasses 50,000 acres of productive area and more than 1,300 wells. Pressure in this unit was maintained by injecting water along the longitudinal axis of the field to create a center-to-edge water-injection pattern. The initial installation for this unit was a single water treatment plant with three associated injection stations. The system was capable of injecting 200,000 B/D water.

There are many examples of single-station water-injection systems, for this is the most common type of water-injection plan. Smaller waterfloods in terms of surface area, number of wells, and water-injection rates are almost universally designed with single injection stations. Even larger operations can be handled with only a few (three to six) injection stations. For example, the Prudhoe Bay field waterflood injection system consists of only six stations injecting more than 2 million B/D water.

On the other hand, the Norge Marchand Unit[29] used a multiple-injection-station system, even though there are only 22 water-injection wells. Six separate water-injection stations serve two to five wells each. Four of the stations are located close to the four wells that supply water for the flood. Very wide variations of permeability and thickness characterize the pay zone at this unit. Generally, wells to the south accepted water with no surface pressure required. Wells to the north had lower permeability and required positive surface pressure for water injection. Because of these variations and the need for immediate injection into the southern wells, multiple injection stations were developed sequentially, starting in the southern part of the unit. Complete waterflood development took more than 1 year.

When waterfloods are planned, the use of multiple water-injection stations should be closely evaluated. Some of the advantages of the multiple system are that (1) failure of equipment or shutdown for repairs does not affect the entire system, (2) shorter distances from injection pumps to injection wells may reduce line costs owing to less length and smaller diameter, (3) injection pressures are flexible for different groups of wells, (4) savings by use of lower total pump horsepower are possible because of reduced friction losses, and (5) flexibility is added for putting produced water into the system when necessary.

One disadvantage may be the lack of centralized operation and data gathering. This can be overcome by modern automation systems that permit data to be gathered and transmitted to a central station. Such systems will add to the initial costs. State-of-the-art supervisory control systems detect problems in the field essentially instantaneously. Personnel can then be dispatched to correct the problem as needed. Data acquisition and retention in these systems are more accurate and more accessibly stored than with manual procedures. In Abu Dhabi's Upper Zakum field,[30] a supervisory control and data acquisition (SCADA) system assists in operating all phases of this 1 million-BOPD field. Westerman[31] gives an excellent discussion of advantages and disadvantages of these systems compared with manual ones. He also covers the recent advances in this ''high-tech'' application to oilfield operations. The extra cost for installing SCADA-type systems must be included in the economic analysis. Another potential disadvantage of multiple-station systems is higher cost per unit of power installed owing to the smaller driver size, as discussed in Sec. 5.14.2.

The number and location of injection stations are important engineering decisions that must be made early in the waterflood planning stage. Each waterflood will have specific conditions that must be taken into account in determining station numbers and locations. The economics, including the cost of initial installation and operations for several possible systems, should be evaluated. The system with the best economics is the one to adopt.

5.17 Associated Facilities

The injection station includes a number of auxiliary systems that supply, service, or control the central pump/driver units. The waterflood engineer's design must include these systems. The cost of these associated facilities is frequently a significant part of the economic analysis.

5.17.1 Surge Tanks. The surge or suction tanks that supply water to the injection pumps are usually built of steel or fiberglass materials with (if possible) a height adequate to satisfy NPSHR needs for the pumps. A tank capacity equal to 5 to 10 hours of injection operation is recommended as a minimum for most waterfloods. This will allow for interruptions from routine maintenance of filters, treating facilities, supply pumps, or other upstream systems supplying water to the tanks so that water injection need not stop. This is especially important in cold climates because of the danger of freezing. Liquid-level controls are frequently necessary in surge tanks to maintain the desired fluid level by starting and stopping transfer or supply pumps and by positioning valves.

5.17.2 Fuel or Electrical System. If the prime mover is fueled by natural gas, a system must be designed to supply that fuel. The system will include piping from the source to the driver units. Desired flow rates and allowable pressure drops in the system must be determined to design the line layout and sizing. If a gas turbine is to be used, a fuel-gas compressor will likely be needed. The fuel system may need to include treating processes such as dehydration or filtration, particularly if casinghead or wet gas is to be used.

Alternative sources of fuel gas, such as propane tanks, may need to be included in the design if the dependability of the primary supply is questionable.

If the energy source is to be electricity, then the engineer may have to determine the sizing and routing of the powerline(s), the primary voltage to be used for transmission, the size and location of transformers and capacitors, and the voltage to be used in the motors. Starting voltage must be taken into account in the supply system as well. Motor-protection devices, shutdowns, and controls must also be designed into the system. Wiring and conduit must be sized and routing determined for the electricity-distribution system within the facility. The electric-system design must meet all applicable codes. Permits and inspections by local regulatory agencies may also be required.

5.17.3 Building and Foundations. The first question to be answered in this area is whether a building is required and, if so, what type. In many climates, either no building or only a roof structure is needed. The engineer should consider such factors as temperature, winds, dust or sand storms, rainfall, snow or ice, and other environmental conditions in determining the need for and design of any buildings. Assuming that a building is needed, the engineer must determine the size, allowing for future growth and the clearance necessary for maintenance and repair activities. Heating and ventilation requirements should be determined. The expected noise level can be estimated and appropriate steps taken to ensure compliance with local and national regulations. Sound baffles or other noise-reduction techniques may be required. The building should be designed to withstand the maximum expected (100-year storm) environmental conditions, particularly wind-loading. Building lighting should be designed to allow adequate light for normal operating and maintenance procedures. The building design should also consider how equipment such as pumps and drivers will be overhauled or replaced. Removal through the walls or roof should be allowed if appropriate. In many cases, it is prudent to provide for, or to install in the original facility, overhead or portable crane units that can be positioned over the equipment. The engineer should also evaluate the type of construction—e.g., prefabricated or built onsite—that is most economical.

Foundations will be required for the building and equipment. The engineer must determine whether slab, pier, or block foundations are optimum for the building and the individual items of equipment. Skid-mounted equipment will usually allow using slab- or rail-type foundations at considerably less expense than the block type. Drain systems for normal washdowns and possible leaks must be included in the design. Environmental discharge regulations should be considered in all drainage designs.

The details of building and foundation design are beyond the scope of this monograph. The waterflood engineer should consult civil engineers or architects as needed, particularly for more complex facilities.

5.17.4 Other Systems. The control systems will be discussed in Chap. 8. Considerations in valve selections include pressure ratings, sizing, materials of construction, pressure drop through the valve, and location. The engineer should ensure that an adequate number of valves are appropriately located in the piping to allow safe operation during maintenance and repair procedures with a minimum of downtime.

In a few cases, regulatory compliance will require exhaust gas or other discharges be controlled to eliminate or to minimize pollution. The engineer should be aware of this possibility, but it will not be dealt with further in this monograph.

5.18 Safety

Many waterfloods concentrate high-pressure equipment in a relatively small area. The facilities often include pressured water and electricity in close proximity. These factors lead to an inherent hazard in waterflood facilities. Design planning must emphasize safety.

5.18.1 High Pressure. The pressures common in modern waterfloods can kill personnel struck by a water stream from a pinhole or flange leak. The engineer must design safeguards into the injection station to include at least (1) safety relief valves and rupture disks with required safety guards as needed; (2) extra wall thickness to allow for corrosion and erosion and design to minimize both; (3) hazard analysis of the piping system to ensure that employees cannot accidently create a dangerous high-presssure condition; and (4) preventive maintenance programs to detect and repair possible problems caused by erosion or corrosion before accidents occur.

5.18.2 Electrical Hazards.
Electricity and water do not mix. If possible, place motor controls in a room where the control center will not be sprayed if there is a leak. In some instances, outside installation may be advisable. The engineer should be sure that the design does not expose personnel to live wires. Electrical equipment should be equipped with lock-out devices to prevent unintentional power activation.

5.18.3 Guards.
All moving parts should be guarded. In the U.S., state and federal regulations dictate this guard, and competent manufacturers will supply them with their equipment. The engineer should inspect fabricated items to ensure that guards are properly installed.

5.18.4 Sour Gas.
The presence of H_2S is common in waterflood operations. Any water containing H_2S will create a hazard during a leak. H_2S is fatal at low concentrations. Because it is heavier than air, this gas will collect in low spots. Thief hatches on surge or supply tanks can become sources of high H_2S concentrations when opened. Vent lines should take H_2S discharges into account to prevent hazards to personnel. Monitoring systems should be installed as appropriate to ensure safe conditions.

5.19 Cold-Weather Considerations

Cold climates affect waterfloods in several obvious ways: (1) freezing water expands and can destroy equipment and piping, (2) cold engines are difficult to start, (3) an employee working in the cold has reduced work efficiency, which approaches zero below 0°F, (4) blowing snow can get into electrical gear and create shorts, and (5) if severe cold (below 0°F) is encountered, equipment failure resulting from steel brittleness can occur. Most of these items indicate the need to house the pumps and drivers with the necessary control valves inside a heated building. In a well-designed waterflood, it is seldom necessary to add heating units to water-storage tanks. However, this should be considered in cold latitudes (above 45°N or 45°S), particularly in continental climates. Insulation on exposed, exterior piping can trap or conduct moisture from blowing snow or precipitation to the pipe wall. Corrosion damage can occur under this insulation, creating a hidden hazard.

5.20 Special Considerations for Offshore Platforms

Waterflooding from platforms has become accepted to the point that it is part of the platform fabrication. Most recent platforms offshore California and many in the North Sea were erected with waterflood equipment in place.

Design for a platform system is simultaneously much more difficult and easier. This is not the paradox it seems. It is easier because many decisions are inherent. For example, the location is on the platform with short lines to the wellheads. The system is harder to design because reservoir data are usually limited until the platform is in place, but the work must be done as part of the platform design. For greatest economic advantage, waterflood equipment should usually be installed on the deck during platform fabrication.

Deck space and weight on a platform are expensive. A first consideration, then, is to design a system with a minimal "footprint." The engineer should give serious consideration to turbine-driven units. Less obvious is the need to evaluate optimal capacity per square foot of deck space used for filters or other treatment facilities. Generally, filters and oxygen-removal equipment will use more deck area than pumps and drivers. In addition, filters are extremely heavy and require satisfactory deck strength. These factors must be considered in the design of both the filters and the platform.

5.20.1 Vibration.
Another consideration on an offshore platform is vibration. The forces of a large, horizontal-plunger pump are not well absorbed by a platform structure. Thus, the great weight and large horizontal forces of a horizontal-plunger pump make this a less desirable selection. A vertical-plunger pump generates forces in a direction that the structure can tolerate. However, this unit still has the disadvantage of a relatively large size and weight. Generally, the optimum unit will be a rotating unit such as a centrifugal pump driven by an electric motor or a turbine. Coincidentally, the fuel cost on many offshore platforms is quite low, thus favoring the compact size and lesser weight of gas-turbine centrifugal-pump packages even though the efficiencies may be low.

5.20.2 Drivers.
An electric motor driver requires that both a generator and the motor be installed on the platform. A direct driver is usually more efficient than the indirectly powered electric motor (particularly at high horsepower). More than onshore, the offshore pump package will tend to be as large as possible, commonly resulting in one pump per platform. This simplifies the piping and controls but requires more forethought to design flexibility.

5.20.3 Retrofits.
Retrofitting on a platform may cost significantly more than an original installation. Designing a waterflood for a platform not originally built with this in mind requires ingenuity. Extra expense usually will be incurred for platform reinforcement and because of the need to cut and rebuild structural members. The cost for on-site welding can be reduced somewhat by prefabrication of welded spools. Electrical, remote control, and utility service must be installed at high offshore costs.

5.20.4 Safety.
Platform waterflood equipment continuously exposes personnel to high-horsepower and high-pressure systems. Considerable extra or redundant controls should be considered. High-temperature turbine exhaust gases must be vented where the employees or mechanical equipment are not adversely affected. Because of the confined spaces and the difficulty of escape, safe design is extremely critical.

5.20.5 Pollution.
Some platforms are located in air quality control areas, such as offshore California, which will require evaluation of controls for the driver exhaust gas. This may require choosing a less optimum prime mover to reduce emissions.

Water pollution can occur from overboard disposal of produced water because of waterflood injection shutdown. This pollution is normally from oil in the water. Another source of potential contamination is discharged treated injection water. Either sea water with chemicals in it or water from a subsea source may violate regulatory standards. These standards often dictate the equipment requirements for treating the water before disposal. The regulations must be considered from the start of platform design.

Nomenclature

C_1, C_2 = constants based on pump type (see Table 5.5)
d = ID of suction piping, in. [cm]
d_{pi} = pump impeller diameter
g = acceleration of gravity, 32.17 ft/sec^2 [9.8 m/s^2]
h = formation thickness, ft [m]
H_a = NPSHA, ft [m]
i_w = water-injection rate, B/D [m^3/d]
k = absolute permeability, md
k_{ro} = relative permeability to oil, fraction
k_{rw} = relative permeability to water, fraction
K = constant based on fluid type, 1.4 for water and 2.5 for hot oil
L = length of suction piping, ft [m]
N = pump speed, rev/min
p_a = pressure exerted at water surface, psi [kPa]
p_{ac} = acceleration pressure losses, psi [kPa]
p_d = pump discharge pressure, psi [kPa]
p_e = pressure at external boundary, psi [kPa]
p_f = flowing frictional pressure losses, psi [kPa]

p_{vp} = vapor pressure of water at system temperature, psi [kPa]

p_w = BHP at injection well, psi [kPa]

p_z = pressure head from surface of water to centerline of pump suction, psi [kPa]

$\Delta p = p_e - p_w$, psi [kPa]

P_{hp} = input power required, hp [kW]

P_{hhp} = hydraulic horsepower, hp [kW]

q = flow rate, gal/min [m³/min]

q_d = pump discharge flow rate, BWPD [m³/d water]

r = radius of external boundary, ft [m]

r_e = outer radius of oil bank, ft [m]

r_w = well radius, ft [m]

v = velocity of fluid, ft/sec [m/s]

γ = specific gravity of liquid

μ_w = water viscosity at reservoir conditions, cp [mPa·s]

μ_o = oil viscosity at reservoir conditions, cp [mPa·s]

References

1. Craig, F.F. Jr.: *The Reservoir Engineering Aspects of Waterflooding,* Monograph Series, SPE, Richardson, TX (1971) **3**, 75.
2. Buckwalter, John F. *et al.*: "Waterflood Oil Recovery Is Lessened By Restricting Rates," *Handbook of Modern Secondary Methods,* Petroleum Publishing Co., Tulsa, OK (1959) 15–26.
3. Howard, G.C. and Fast, C.R.: *Hydraulic Fracturing,* Monograph Series, SPE, Richardson, TX (1970) **2**.
4. Preuninger, P.D.: "Motor Selection For Special Environments," Louis Allis Div., Allis-Chalmers Corp., Milwaukee, WI, undated.
5. "Standard Practices for Low and Medium Speed Stationary Diesel and Gas Engines," fifth edition, Diesel Engine Manufacturers Assn., Washington, DC (1958).
6. *Standard 7B-11C, Specification for Internal-Combustion Reciprocating Engines for Oil Field Service,* eighth edition, API, Dallas (March 1981).
7. Moe, R.C.: "Engine Sizing, Running Practices Lead To Diesel Savings," *Oil & Gas J.* (March 23, 1981) 125–31.
8. *Engineering Data Book,* ninth edition, Gas Processors Assn., Tulsa, OK (1972).
9. *Industrial Gas Engines: Performance Curves and Technical Information,* Caterpillar Tractor Co., Peoria, IL (June 1982).
10. Marquis, D.E.: "Gas Engines Offer Potential for Energy Savings," *Oil & Gas J.* (Dec. 29, 1980) 191–98.
11. Neal, J.C.: "Gas Turbine Driven Centrifugal Pumps for High-Pressure Water Injection," paper SPE 1888 presented at the 1967 SPE Annual Meeting, Houston, Oct. 1–4.
12. Woodhall, R.J.: "Turbines Power Cook Inlet Waterflood," *Pet. Eng.* (July 1970) 56–62.
13. Moore, B.L. Jr.: "Webster Field Unit Waterflood Facilities," paper SPE 6880 presented at the 1977 SPE Annual Technical Conference and Exhibition, Denver, Oct. 9–12.
14. Taylor, I.: "NPSH Still Pump-application Problem," *Oil & Gas J.* (Jan. 2, 1978) 103–08.
15. Miller, J.E.: "Liquid Dynamics of Reciprocating Pumps—1," *Oil & Gas J.* (April 18, 1983) 91–112.
16. Karassik, I.J.: "Cavitation and Its Effect on Centrifugal Pump Operation," *Chemical Processing* (Dec. 1981) 18–21.
17. Henshaw, T.L.: "Reciprocating Pumps," *Chem. Eng.* (Sept. 21, 1981) 105–40.
18. *Marks' Standard Handbook for Mechanical Engineers,* T. Baumeister (ed.), eighth edition, McGraw-Hill Book Co., New York City (1979).
19. Miller, J.E.: "Liquid Dynamics of Reciprocating Pumps—2," *Oil & Gas J.* (May 2, 1983) 180–93.
20. Langewis, C. Jr. and Gleeson, C.W.: "Practical Hydraulics of Positive Displacement Pumps for High-Pressure Waterflood Installations," *JPT* (Feb. 1971) 173–80.
21. Karassik, I.J.: *Centrifugal Pump Clinic,* Marcel Dekker Inc., New York City (1981).
22. "Selection of Metallic Materials To Be Used in All Phases of Water Handling for Injection into Oil Bearing Formations," RP-04-75, Natl. Assn. of Corrosion Engineers, Katy, TX (April 1975).
23. Holland, A.L. Jr.: "Water-Plant Success by Design," *Pet. Eng.* (Nov. 1983) 108–20.
24. Johnson, J.L. and Vandevier, J.E.: "Horizontal Pumps: A New Approach to Water-Injection-Pressure Boosting," paper SPE 14260 presented at the 1985 SPE Annual Technical Conference and Exhibition, Las Vegas, Sept. 22–25.
25. Stevenson, J.M.: "Gas Engines or Electric Motors," *Pet. Eng.* (April 1961) B/19-25.
26. Blanton, M.L.: "Shell Extends Ventura Flood to 5000 psi," *Pet. Eng.* (Aug. 1974) 24–26.
27. Moore, H.J.: "A Tailor-Made Water-Injection System Saves Money in the LL-5 Flank Waterflood at Lake Maracaibo," *JPT* (Dec. 1961) 1191–94.
28. Thomson, J.A., Emerick, G.M., and Kostrop, J.E.: "Integration/Coordination: Keys to SACROC Field Operations," *Pet. Eng.* (Dec. 1971) 48–56.
29. Jowell, J.H. and Haburther, C.R.: "Norge Marchand Deep Injection Project," *Pet. Eng.* (Nov. 1976) 23–33.
30. "New SCADA Technology Used to Operate Upper Zakum Field," *Pipeline Ind.* (Oct. 1984) 26–27.
31. Westerman, G.W.: "Improved Controls for Enhanced Recovery," *Proc.,* 29th Annual Southwestern Petroleum Short Course, Lubbock (1982) 339–348.

SI Metric Conversion Factors

acre × 4.046 873	E−01	= ha
acre-ft × 1.233 489	E+03	= m³
bbl × 1.589 873	E−01	= m³
Btu × 1.055 056	E+00	= kJ
Btu/ft³ × 3.725 895	E+01	= kJ/m³
Btu/(hp-hr) × 3.930 148	E−01	= W/kW
cycles/sec × 1.0*	E+00	= Hz
ft × 3.048*	E−01	= m
ft³ × 2.831 685	E−02	= m³
°F (°F−32)/1.8		= °C
hp × 7.460 43	E−01	= kW
in. × 2.54*	E+00	= cm
in.² × 6.451 6*	E+00	= cm²
lbm × 4.535 924	E−01	= kg
psi × 6.894 757	E+00	= kPa

*Conversion factor is exact.

Chapter 6
Injection System

6.1 Introduction

In Chap. 5, injection-pump design discharge pressure was calculated assuming a pressure drop in the distribution lines of 100 psi [690 kPa] or 10% of the discharge pressure, whichever was greater. A further assumption was a pressure drop in the wellhead and downhole assembly of about 50 psi [345 kPa]. These simplifying assumptions allow pump and prime-mover selection to be made so that station design can proceed. This chapter discusses a more accurate method of sizing injection lines and also briefly covers injection-well design.

6.2 Designing Injection Piping Systems

The design of an injection piping system can be divided into two segments: discharge piping in the plant and field distribution lines. The former is primarily a function of the number and type of pumps, while the latter is determined by the economic tradeoff between line size and required horsepower. The design of both types of systems needs to be flexible to allow for future changes in the waterflood. As a flood develops and matures, the location of water-injection wells and the rates injected by them will likely change from the initial conditions. The basic cause for such variances is the difference between actual and expected reservoir performance. This difference may result in new infill drilling, changes in the pattern type, redirection of the available water injection volumes, or an increase/decrease in injection volume. In turn, such modifications will affect the correct choice of line sizes, lengths, materials, pressure ratings, and similar considerations. The engineer must evaluate the probable future changes and design the system to accommodate them in the most economical manner. For example, if increasing the rates is a fair probability, then using a 12-in. [30-cm] -diameter suction manifold when an 8-in. [20-cm] pipe is calculated to be sufficient is a good practice. The cost of doing so is negligible and may result in a large future savings in retrofitting costs and shutdown time. The engineer must weigh the certainty of an initial cost for flexibility against the chance of future retrofitting costs.

6.2.1 Discharge Piping Design.
Discharge piping design requires some special features. At higher pressures (particularly with the use of plunger pumps), it is important to follow correct principles. Sharp bends, valves, reducers, or other causes of turbulence should be limited as much as possible. The piping near the pumps should be sized to conduct water at velocities equal to or exceeding 3 to 4 ft/sec [0.9 to 1.2 m/s]. For positive-displacement (PD) pumps, a pressure-relief valve should be located ahead of any other valve or obstruction. Elbows or other fittings should not be located at or near the pump unless absolutely necessary. All piping near the pump should be designed to provide as smooth a flow path as possible.[1,2] Any required elbows should be the long-radius type or even fabricated sweeps. If possible, "tees" should be avoided in manifolds.[3] Fabricated sweep pieces are usually cost-effective and preferable to elbows and tees (Fig. 6.1).

The injection discharge piping must be sized to have the necessary pressure rating. Knowing the design discharge pressure, the engineer can consult the latest editions of ANSI Standards B31.3,

B31.4, B31.4b, or B31.8 for piping and field-line design guidelines. Tables 1 and 2 summarize pressure ratings by weight per foot of wall thickness for the more common standard carbon-steel piping sizes per ANSI Standard B31.4. Fittings, valves, meters, and other accessories must also be specified to meet or exceed the pipe pressure rating. Some allowance for corrosion should be made for bare steel lines. Various allowances can be chosen, depending on whether the expected corrosivity of the water is low or moderate. Allowances of $\frac{1}{16}$ or 0.05 in. [1.6 or 1.3 mm] are commonly used for moderate corrosivity. If severe conditions are expected, the lines should be protected (see Sec. 6.2.3). In most waterfloods, Schedule 40 pipe and standard weight fittings are adequate. Where replacement and repair costs are high (offshore and arctic environments), Schedule 80 or even heavier pipe is often used to increase service life and reliability.

When a bypass valve and line is provided to start a PD pump under no-load conditions, the bypass line should be at least half the size of the discharge line. The bypass should be piped into a supply tank rather than directly back to the suction line. Sec. 5.13.4 covers bypass design and controls for turbine-driven, centrifugal pumps. A valve should be provided to prevent flowback through the pump while it is being serviced. Discharge and suction lines should be rigidly anchored and supported so that the pipe is not dependent for support on the machinery to which it is connected. Langewis and Gleeson[3] discuss a calculation of spacing for pipe supports based on the concept of the highest estimated pulsation frequency:

$$F_p = \frac{n \times N \times n_p}{60}, \qquad \qquad \qquad (6.1)$$

where
F_p = highest pulsation frequency,
n = number of plungers in pump,
N = pump speed, and
n_p = number of pumps on manifold.

Then the piping frequency, f_p, is assumed at $4 \times F_p$ and the spacing can be calculated by

$$\Delta L = [(\alpha / f_p) \times (E \times I / W)^{0.5}]^{0.5}, \qquad \qquad (6.2)$$

where
ΔL = pipe support spacing,
α = coefficient = 1.69 for pipe fixed at both ends,
f_p = $4 \times F_p$ (Eq. 6.1),
E = modulus of elasticity, 30×10^6 lbf/in.^{2} [133×10^6 N/cm^2] for carbon-steel pipe,
I = moment of inertia (Table 2), and
W = pipe weight.

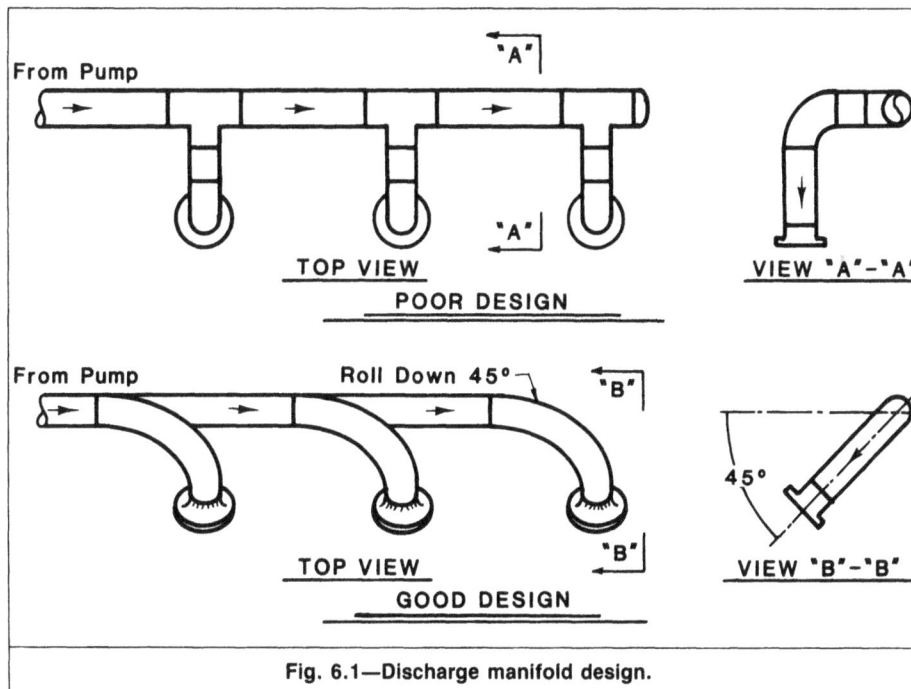

Fig. 6.1—Discharge manifold design.

Langewis and Gleeson suggest randomly varying the actual spacing a little from the calculated value to prevent synchronous vibrations. A variance of 0 to 10% should be sufficient. An example of support spacing calculations is shown in Appendix B.

At times, it is necessary to install pulsation dampeners or devices to reduce pulsations in the system. Two basic types of dampeners are the bladder type and the flow-through, acoustical type. Sec. 5.11.2 discusses these devices in more detail. The in-line bladder-type dampener has proved to be more effective than the appendage bladder type.[3] The dampeners must be sized to meet individual pump requirements. Good piping design will minimize or eliminate the need for such dampeners. In many offshore and some onshore installations, the room for the injection system is minimal. This limitation may dictate the use of pulsation dampeners because the piping cannot be adequately designed and built in the available space. If the engineer is uncertain about the requirements, the piping can be designed with a straight run of pipe near the discharge flange of enough length to allow installation of a dampener later.

Pressure drop in the discharge piping can be calculated with the concepts of equivalent lengths or resistance coefficients for the loss through fittings. Several manufacturers of fittings have published information on these calculation techniques.[4,5] The waterflood engineer is encouraged to consult such documents. The concept of equivalent lengths is simply that each fitting or valve will have a pressure drop equal to the drop across a certain length of pipe. This length has been calculated and is contained in Refs. 4 and 5. For example, a 6-in. [15-cm] nominal, long-radius, 90° [1.6-rad] elbow has an equivalent length of 6.5 ft [2 m]; i.e., the pressure drop across the elbow is the same as the reduction over 6.5 ft [2 m] of 6-in. [15-cm] nominal pipe. In calculating the total pressure drop, the piping length should be increased by 6.5 ft [2 m] to account for the presence of the elbow. For a given piping run, the total equivalent lengths of all the valves and fittings must be added to the piping length to calculate the pressure losses in the system. The total pressure drop in the discharge piping should be from a few psi to several tens of psi, depending on the system pressure.

6.2.2 Injection-Distribution-System Design. Distribution-system design is basically a tradeoff between the cost of larger-size pipe and the cost of additional horsepower at the injection station. As discussed in Sec. 5.16, the injection-line design is a function of the number and location of injection stations. In many cases, this will be determined by the location of existing facilities or other factors. Where large, new fields are being developed, an algorithm to optimize water-injection lines as well as other facilities is avail-

able. The procedure outlined by Mathur et al.[6] is a trial-and-error process, as are all other calculations of this nature. Valaster et al.[7] also developed a mathematical model that uses numerical solutions to calculate pressure drops and line sizing. Generally, the engineer can design systems more simply, but these procedures should be useful for complex systems. In most cases, the engineer will find that injection-line costs will influence, but not determine, the number and location of injection stations. Some companies have developed or acquired computer programs to optimize these design calculations. The engineer should check on the availability and applicability of these programs.

Two basic types of water injection distribution systems exist: radial and trunkline. The radial system usually is used when a limited number of injection wells (up to about 12 to 15) will be used for any one station. When a large number (>15) of injection wells are to be served by one injection station, the trunkline system is generally more economical. Combinations of these two basic types are often a good design solution. For example, trunklines from a central plant to several satellite distribution manifolds with radial lines to the wells may be the most economical choice for a given field. At Prudhoe Bay, the main trunklines from the injection plants go to the drillsites, where a distribution manifold directs the water to individual lines to the wells. In this field, the central injection plants are all connected to increase flexibility and to protect against frozen lines.

The shape of the waterflood area will help to determine the most desirable type of system. For example, a long, narrow reservoir would be suited for a trunkline along the longitudinal axis with short laterals. A roughly circular reservoir of limited size might be served best by a radial system. A large, circular field might be broken into four or more interconnected injection stations with radial lines or the trunkline system used with a single station. Injection wells are sometimes located such that two parallel trunklines are used.[8] Appendix B shows an example in which both radial and trunkline systems are appropriate.

The type of waterflood pattern selected also will influence the selection of the distribution system. For continuous patterns, such as the five-spot, a trunkline with laterals is usually the best choice. A peripheral injection pattern could use a line that would loop the entire field, serving each injection well. Irregular location of injection wells could be served by a radial system or by several injection stations, such as in the Norge Marchand Unit.[9]

Keeping these principles in mind, the engineer should now calculate line sizes and costs for each system that appear feasible. This trial-and-error process consists of calculating the total pressure drop

TABLE 6.1—PRESSURE RATINGS*

Pipe Size (in.) Nominal	OD	ID	Weight (lbm/ft)	Wall Thickness (in.)	Pressure Rating (psi)	Classification
½	0.84	0.622	0.85	0.109	6,540	Schedule 40 or standard
		0.546	1.09	0.147	8,820	Schedule 80 or XS
		0.466	1.30	0.187	11,220	Schedule 160
¾	1.05	0.824	1.13	0.113	5,424	Schedule 40 or standard
		0.742	1.47	0.154	7,392	Schedule 80 or XS
		0.614	1.94	0.218	10,464	Schedule 160
1	1.315	1.049	1.68	0.133	5,097	Schedule 40 or standard
		0.957	2.17	0.179	6,861	Schedule 80 or XS
		0.815	2.84	0.250	9,582	Schedule 160
1¼	1.66	1.38	2.27	0.14	4,251	Schedule 40 or standard
		1.278	3.0	0.191	5,799	Schedule 80 or XS
		1.160	3.77	0.25	7,590	Schedule 160
1½	1.9	1.61	2.72	0.145	3,846	Schedule 40 or standard
		1.5	3.63	0.2	5,305	Schedule 80 or XS
		1.338	4.87	0.281	7,454	Schedule 160
2	2.375	2.067	3.65	0.154	3,268	Schedule 40 or standard
		1.939	5.02	0.218	4,626	Schedule 80 or XS
		1.689	7.46	0.343	7,279	Schedule 160
2½	2.875	2.469	5.79	0.203	3,559	Schedule 40 or standard
		2.323	7.66	0.276	4,838	Schedule 80 or XS
		2.125	10.01	0.375	6,574	Schedule 160
3	3.5	3.068	7.7	0.216	3,110	Schedule 40 or standard
		2.9	10.33	0.3	4,320	Schedule 80 or XS
		2.624	14.31	0.438	6,307	Schedule 160
4	4.5	4.026	10.79	0.237	2,654	Schedule 40 or standard
		3.826	14.98	0.337	3,774	Schedule 80 or XS
		3.438	22.52	0.531	5,947	Schedule 160
6	6.625	6.065	18.97	0.28	2,130	Schedule 40 or standard
		5.76	28.57	0.432	3,286	Schedule 80 or XS
		5.187	45.34	0.719	5,470	Schedule 160
8	8.625	8.125	22.36	0.25	1,461	Schedule 20
		7.981	28.55	0.322	1,882	Schedule 40
		7.625	43.39	0.5	2,922	Schedule 80 or XS
		6.813	74.71	0.906	5,294	Schedule 160
10	10.75	10.25	28.04	0.25	1,172	Schedule 20
		10.02	40.48	0.365	1,711	Schedule 40
		9.75	54.74	0.5	2,344	Schedule 60 or XS
		9.562	64.4	0.594	2,785	Schedule 80
12	12.75	12.25	33.38	0.25	988	Schedule 20
		11.938	53.56	0.406	1,605	Schedule 40
		11.626	73.22	0.562	2,222	Schedule 60
		11.374	88.57	0.688	2,720	Schedule 80
14	14.0	13.376	45.68	0.312	1,123	Schedule 20
		13.124	63.37	0.438	1,577	Schedule 40
		12.812	85.01	0.594	2,138	Schedule 60
		12.5	106.13	0.75	2,700	Schedule 80
16	16.0	15.376	52.36	0.312	983	Schedule 20
		15.0	82.77	0.5	1,575	Schedule 40 or XS
		14.312	136.58	0.844	2,659	Schedule 80

Temperature and Corrosion Allowance Factors

Temperature (°F)	Correction Factor	Nominal Pipe Size (in.)	Δp/0.01 in. (psi)
<250	1.0	1	383
300	0.967	1¼	304
350	0.933	1½	265
400	0.9	2	212
450	0.867	2½	175
		3	144
		4	112
		6	76
		8	58

*Information is from ANSI Standards B31.4 (1979) and B31.8 (1982) for plain-end, seamless or ERW, Grade B, carbon-steel line pipe according to API Standard 5L. The temperature range is −20 to 250°F.

in a given system assuming a pipe size. If the drop is excessive—i.e., exceeding the larger of 100 psi [690 kPa] or 10% of the maximum system pressure—then the line size(s) is (are) incremented upward and the process repeated. Pressure drops can be calculated from the Darcy (Fanning) equation using the Moody friction factor, f:

$$\Delta p = 0.000216 f L \rho_w q_w^2 / d^5, \qquad \ldots\ldots\ldots\ldots\ldots (6.3)$$

where

Δp = pressure drop,
f = Moody friction factor,
L = length of pipe,
ρ_w = water density,
q_w = water flow rate, and
d = ID of pipe.

Another equation for pressure drop with the flow of water in pipes is the empirically derived Hazen-Williams formula:

$$v = C \times r^{0.63} \times (H_f/L)^{0.54} \times 1.32, \qquad \ldots\ldots\ldots\ldots (6.4)$$

where

v = velocity,
C = Hazen-Williams coefficient,
r = hydraulic radius, and
H_f = friction head,

which can be rearranged as

$$\Delta p = L \times [q_w/(15.1889 \times C \times d^{2.63})]^{1/0.54} + 0.4331 \Delta h, \quad \ldots (6.5)$$

where

L = pipe length,
q_w = water flow rate, and
Δh = elevation change of pipe.

Recent work[10] has shown that, for modern steel pipe finishes, a coefficient of 140 is more correct than the traditional value of 100, which will lead to a more conservative design. When the Hazen-Williams formula is used, the engineer should estimate pump and line sizes with both values of the coefficient. This estimate will determine the economic consequences of using the more conservative design and allow the engineer to make the proper choice of equipment. Bartos[11] developed solutions for Eq. 6.5 on programmable calculators. Most companies have programs available on their computers to solve for pressure drops in piping. The engineer is advised to use these programs to speed the solution of these trial-and-error procedures.

In most cases, calculation of only one or two flow paths will give the engineer a solution for the maximum pressure drop. Even in complex systems like the Prudhoe Bay and Kuparuk waterfloods, only a few flow paths had to be analyzed to ensure that the maximum pressure drop was found. Obviously, the engineer should look for the longest length of line from pump to wellhead with the highest flow rate. Frequently, this can be done by inspection of the data. Once appropriate line sizing that meets the pressure-drop criteria has been determined for this (these) segment(s), the rest of the system can be calculated fairly quickly and laid out by the waterflood engineer. Appendix B shows an example problem worked out for a distribution system including piping pressure drops. The engineer is cautioned to change line sizes (when required) at or near wells as much as possible. This design feature will facilitate the addition of pig launchers and traps later in the waterflood. In many cases, a buildup of scale or corrosion products necessitates cleaning of the injection lines with pigs and scrapers.

Once acceptable pressure drops are calculated for a given set of pipe sizes in an injection system, the velocity of the water flow should be calculated. The velocity of the water in a pipe is equal to the flow rate divided by the cross-sectional area of the pipe. For circular pipe,

$$v = \frac{q_w}{A} = q_w/(\pi r^2) = 0.408 q_w/d^2, \qquad \ldots\ldots\ldots\ldots (6.6)$$

TABLE 6.2—MOMENT OF INERTIA			
Nominal Pipe Size (in.)	Moment of inertia (in.^{4})		
	Schedule 40	Schedule 80	Schedule 160
1	0.0874	0.1055	0.1252
1½	0.310	0.391	0.483
2	0.666	0.868	1.163
3	3.02	3.90	5.04
4	7.23	9.61	13.27
6	28.1	40.5	59.0
8	72.5	105.7	165.9
10	160.8	244.8	399.4
12	300	475	781
14	429	687	1,117
16	732	1,157	1,894
18	1,171	1,834	3,020
20	1,704	2,772	4,585
24	3,420	5,670	9,455

where v = velocity. Maximum allowable velocity in the chosen steel piping system should not exceed about 8 ft/sec [2.4 m/s] . Velocities of 10 ft/sec [3 m/s] have been cited as accelerating corrosion/erosion in steel pipes.[12] A minimum velocity of about 3 ft/sec [0.9 m/s] is recommended to keep the line clean. When all feasible systems have been quantified in terms of cost, the most economical choice can be made. Cost factors should be available to the waterflood engineer from recent installations, local production supervisors, other engineers, or vendors. While these may not be of the necessary quality for a final expenditure-approval document, such estimates will allow choosing the lowest-cost system.

6.2.3 Corrosion Protection in Injection Lines. Corrosion protection in general is covered in Sec. 4.4.6. This section briefly covers the use of protective coatings in injection systems.

Most waterfloods are designed to provide corrosion control in the water-injection lines by the use of internally coated pipe or fiberglass-reinforced plastic (FRP) pipe. The internal coatings are usually either baked-on plastic or cement. Larger-diameter pipe is often cement-lined because of the lower cost.[13] The available plastic lining materials include powder epoxies, epoxy phenolics, epoxy cresol novolac, and urethane-modified epoxy. The state of the art in plastic linings changes constantly. The waterflood engineer should consult with people knowledgeable in this area before choosing a coating. A reasonable amount of care must be exercised when handling internally coated pipe, particularly cement-lined pipe, to prevent damage to the coating during transportation and installation. Because of the potential for formation damage caused by cement particles sloughing off into the injection water, cement-lined pipe is not recommended when the reservoir permeability is low.

A suitable pipe-joining method must be selected to maintain the coating integrity. The engineer should obtain operating information on the joining methods available in an area. If necessary, other experienced operators or coating contractors can be consulted.

The API has published a recommended practice[14] that is a good guideline to cement lining of steel pipe. Quality control during the coating process is critical to the success of the corrosion protection. Generally, this is an area in which the waterflood engineer should get advice from purchasing agents, company experts, or consultants.

There are waterfloods where no corrosion protection other than chemical inhibitors is required and bare steel pipe is used. If the flooding water is not corrosive or only mildly so, or if the waterflood is expected to last a short period of time, economics may dictate that less expensive bare pipe be used. When there is a question of whether bare pipe should be used, waterflood operators sometimes use bare pipe, anticipating the possible replacement of some or all of the pipe in the later years of the flood. This practice is not recommended in most cases because of the cost of failure repairs and plugging of wells caused by corrosion products. Some waterfloods have been abandoned early because of the high cost of replacing pipe or other corroded equipment. If there is any doubt about the degree of corrosion that may be experienced in the future, the

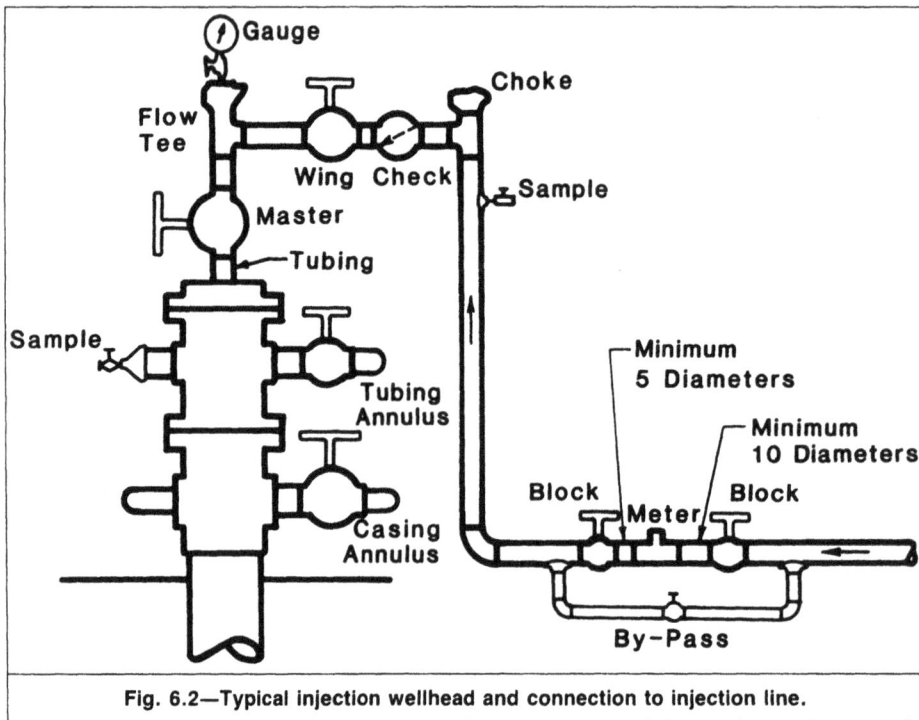

Fig. 6.2—Typical injection wellhead and connection to injection line.

correct decision is to protect the system by chemical inhibition, coatings, and the proper material selection.

It is recommended that FRP pipe be protected from direct sun rays. Therefore, it should be painted or buried unless an effective protective coating is supplied by the manufacturer. For high pressure ratings, this type of pipe is more expensive than coated steel pipe but may be the best choice in severely corrosive waters. Because of the potential for catastrophic failures of FRP pipe (mentioned in Chap. 4), physical protection should be provided to prevent access by personnel or animals where such pipe is used in high-pressure service above ground.

External coatings and wrapping of pipe are required where steel pipe is placed in a corrosive environment. Wet soil conditions or the presence of brine or mud pits may create conditions causing external corrosion. Commonly, buried steel lines are externally coated with hot asphalt or coal-tar enamels. These coatings usually are wrapped with fiberglass or felt banding. Various types of tapes also are used to wrap pipe spirally by hand or by use of wrapping machines in the field. [15] As discussed in Chap. 4, cathodic protection with impressed currents or sacrificial anode beds is frequently economical and will help prevent failures owing to corrosion of buried steel lines.

6.3 Wells

The detailed design of wellhead and downhole assemblies is beyond the scope of this monograph. This section attempts to cover only some fundamentals and guidelines. The waterflood engineer will likely have only limited input into the design of this equipment. Drilling and/or production engineering personnel will normally provide this type of information, but the waterflood engineer may be required to specify it.

6.3.1 Wellhead Equipment. Generally, wellhead equipment, such as casing heads and casing valves, is the same for water-injection wells as for other wells. The casing heads should be provided with one or two side outlets, at least one of which should be equipped with a valve. If the engineer has no other information, then the casing heads being used on nearby production wells will likely be more than adequate for injection service, although corrosion control may be needed.

The tubing-head and Christmas-tree configurations should be kept as simple as is practical. Tubing heads should have at least one outlet equipped with a valve to allow tests for casing integrity, packer/tubing leaks, or other regulatory needs. Tubing hangers can be either

slip type or screw-on type. The slip-type hanger can be set with 2 ft [0.6 m] or so of tubing protruding above the tubing-head flange and with a master valve screwed onto the top. The inlet to the wellhead (valve, tee, or whatever) is then attached to the master valve and the injection line piped up to it. A properly sized check valve should be installed in the line near the wing valve to prevent backflow from the well in the event of a line break, system shutdown, or similar occurrences. Fig. 6.2 shows a typical wellhead design, but any number of variations are possible and virtually all work satisfactorily. Several manufacturers make complete wellhead injection assemblies, including master valve, flow tee with cap, choke assembly, and wing valve. This type of equipment is frequently a good and simple choice for an injection wellhead. The engineer should always provide for adequate sample and pressure taps and temperature wells at or near the wellhead. The wellhead and associated valves should be designed for the appropriate pressure rating. The flow tee, nipples, and master valve should be a minimum of 2 in. [5 cm] ID when logging surveys are anticipated. If logging will occur frequently, a ball valve installed on top of the flow tee will facilitate the operations.

Control equipment is simple in most waterflood injection wells. A meter with a totalizer and a positive choke are frequently all that is required. In some cases, the well is allowed to "ride the line pressure" and inject as much water as possible. In such cases, no flow-control device is needed. Flow-control valves[16] can also be used with or without actuators and sensors to adjust flow rates or pressures automatically. Even more complex systems have been used, such as a solar-powered system[17] that senses the bottomhole pressure and adjusts a control valve to keep it about 50 psi [344 kPa] below fracture pressure. The more complex systems are discussed further in Chap. 8.

All control equipment and the wellhead must be constructed of suitable materials for the service. For reference, the Natl. Assn. of Corrosion Engineers[18] has published a recommended practice on the selection of materials for water injection. Wellheads can be made of carbon steel and internally plastic coated for corrosion protection. Alternatively, the injection wellhead assembly can be made from corrosion-resistant materials such as aluminum bronze or stainless steel.[19] The choke or flow-control valve should be resistant to erosion and corrosion, with stainless steel, chrome, or carbide used as needed. Table 6.3[20] gives a brief description characterizing different metals for these purposes.

Cold winter areas may need special attention in the line and wellhead design. The lines should be buried below the frost line, which may be 4 to 5 ft [1.2 to 1.5 m] deep or more in the northern U.S.

INJECTION SYSTEM

INJECTION SYSTEM
51

TABLE 6.3—MATERIAL RESISTANCE TO CORROSION AND EROSION

Material	Corrosion/Erosion-Resistant Properties
Carbon steel	No resistance to CO_2, H_2S, or oxygen and minimal resistance to salt water. Can be hardened to handle mild abrasives or coated with corrosion/abrasion-resistant materials.
Alloy steel	Can be hardened for good resistance to abrasives. Some alloys are fairly resistant to corrosives, and plating or spraying the surfaces can offer further abrasion resistance.
Stainless steel	No resistance to abrasion unless wearing surfaces are coated with resistant material. Can be alloyed to resist either H_2S or CO_2 corrosion, but may pit when salt water is present.
Brass	Poor abrasion resistance unless plated with chrome or similar substance. Can be alloyed against salt water or CO_2 corrosion and some alloys offer H_2S resistance.
Monel	Poor abrasion resistance unless hard-surface coated. Excellent resistance to corrosion.
Cast iron	Fairly resistant to abrasion. Nickel in cast iron alloys provides moderate corrosion resistance.
Cobalt alloys	Highly resistant to abrasion and excellent resistance to corrosion.
Carbides	Highly resistant to both abrasion and corrosion.

High-alloy coatings are available that have more abrasion and corrosion resistance than low-alloy steels. Care must be taken with chrome plating because chrome is porous. The use of chrome-against-chrome sliding components is not generally favorable.

Materials must be galvanically compatible and attached components should be made of the same or similar materials.

In arctic regions, where the permafrost extends more than 1,000 ft [305 m] deep, the lines should be elevated and insulated. As long as a reasonable flow rate is maintained, the wellheads and aboveground lines will not freeze. Insulation will also help maintain the temperature above freezing. Depending on the length and severity of the expected cold weather, freeze-protection systems may or may not be needed. If electricity is available, installing heat tape on the wellhead under insulation may be a wise precaution against freezing. Multiple pump systems and interconnected injection stations that allow continued injection at reduced rates when pump failures occur may be another viable solution. Because each waterflood is different, the waterflood engineer must evaluate the need for and cost of alternative freeze-protection systems on a case-by-case basis. In cooler climates, a good design practice is to provide a means to drain the lines in the event of a relatively long downtime.

Engineering control of a waterflood requires that injected water be metered. Metering usually is done at the plant where water is pressured and at the individual water-injection wells. Further discussion of metering at the plant is included in Chap. 8.

At the individual injection wells, PD meters are frequently chosen to supply information on individual-well injection rates. These meters may be read manually or impulse counters can read and transmit the data to a central location. More recently, turbine meters have become commonly used and vortex meters have been installed in some cases. These meters have the advantage of handling a wide range of rates for any given size. With the proper read-out equipment installed or available, these meters will provide instantaneous flow-rate data, which facilitate running various pressure/rate tests. While they are slightly more costly, turbine meters are more accurate and fit in better with automation systems than do PD meters. In choosing meter types, the engineer should carefully weigh the need for accuracy against cost. A design should not specify extremely accurate metering (e.g., 0.1%) and repeatability when such accuracy is not justified by the waterflood needs because this practice incurs unnecessary costs. Sometimes the meters for individual injection wells are grouped at an injection-discharge manifold for systems, with individual lines to the wells.

6.3.2 Downhole Equipment. The tubular goods for injection wells are generally bare carbon steel, internally plastic-coated (IPC) carbon steel, or FRP tubing. The same limits and cautions apply to this tubing as were discussed in Sec. 6.2.3. The most commonly used injection tubing is probably one made of IPC steel materials. Coupled with a chemical corrosion-inhibition program, the IPC steel tubing will usually give an adequate service life. In waterfloods with low corrosivity, bare steel tubing may be the most economical. FRP tubing is normally used only for very corrosive environments. The waterflood engineer must again evaluate the conditions of a specific flood to determine what material is most economical.

As a general rule, the more simple a downhole assembly is, the better it is. In most cases, the injection well need be equipped with only a tubing string and a packer (Fig. 6.3). Setting the packer 100 to 150 ft [30 to 46 m] above the top perforation will make logging easier because this allows faster and more accurate collar locating and less turbulence, and helps locate channels behind the casing. However, local or national regulations may dictate the packer setting depth. Sometimes multiple packers and tubing strings are used to provide injection control in several different zones. The forces and stresses associated with tubing/packer assemblies have been addressed in several papers[21,22] and will not be covered in this monograph. Careful calculations of setting forces or forces from thermal and pressure changes resulting from starting or stopping injection must be made to ensure that the tubing is not damaged.

In wells with steel tubing, the downhole injection assembly can frequently use simple, inexpensive, tension-set retrievable packers, particularly in shallow wells. In cases where multiple packers are needed, combinations of permanent or retrievable, hydraulically or mechanically set packers can be used. In more complex completions, the waterflood engineer should obtain recommendations on the downhole assembly from in-house or outside consultants with expertise in this area.

For wells with FRP tubing, packer selection becomes more limited. Technology in manufacturing FRP tubes is improving constantly but, currently, FRP tubing is limited to about 10,000 ft [3050 m] of depth.[23] The tubing-string weight (typically about one-third that of steel) and strength, particularly in shallow wells, are too low

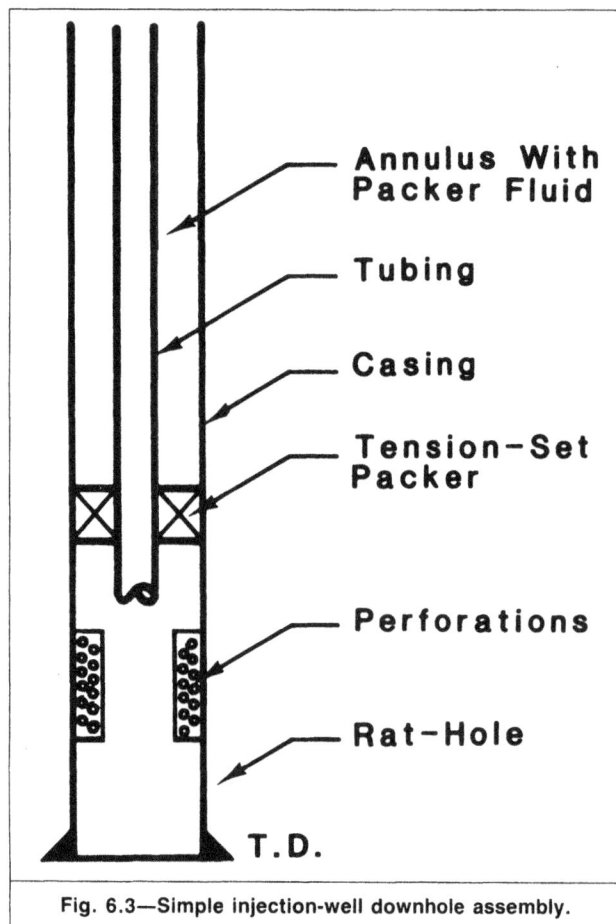

Fig. 6.3—Simple injection-well downhole assembly.

to allow the use of mechanically set packers. Pressure rating will frequently preclude the use of hydraulically set packers. Generally, permanent packers with a seal assembly attached to the bottom of the FRP tubing are the best selection.[23]

The casing/tubing annulus of the injection well should be protected by filling with a noncorrosive packer fluid. This fluid can be oil or water and is usually treated with a corrosion-inhibiting chemical and a biocide. When required, an oxygen scavenger or other chemicals may also be added to the packer fluid. This fluid should be circulated through the annulus to remove all wellbore liquids before the packer or packer seal assembly is set.

Nomenclature

C = Hazen-Williams coefficient
d = pipe ID, in. [cm]
E = modulus of elasticity = 30×10^6 lbf/in.^{2} [133×10^6 N/cm^2] for carbon-steel pipe
f = Moody friction factor
f_p = $4 \times F_p$ (Eq. 6.1), pulses/sec
F_p = highest pulsation frequency, pulses/sec
Δh = elevation change of pipe, ft [m]
H_f = friction head, ft [m]
I = moment of inertia (Table 2), in.^{4} [cm^4]
L = pipe length, ft [m]
ΔL = pipe support spacing, ft [m]
n = number of plungers in pump
n_p = number of pumps on manifold
N = pump speed, rev/min
Δp = pressure drop, psi [kPa]
q_w = water flow rate, gal/min [m^3/min]
r = hydraulic radius, in. [cm]
v = velocity, ft/sec [m/s]
W = pipe weight, lbm/ft [kg/m]

α = coefficient = 1.69 for pipe fixed at both ends
ρ_w = water density, lbm/ft^3 [kg/m^3]

References

1. Haberthur, C.R.: "Improve Profits Through Engineered Pump Station Design," *Proc.*, 20th Annual Southwestern Petroleum Short Course, Lubbock, TX (April 1973) 183–87.
2. Wurzbach, W.M. Jr.: "Measurement of Suction and Discharge Pressure Pulsations in Waterflood Facilities," paper SPE 12194 presented at the 1983 SPE Annual Technical Conference and Exhibition, San Francisco, Oct. 5–8.
3. Langewis, C. Jr. and Gleeson, C.W.: "Practical Hydraulics of Positive Displacement Pumps for High-Pressure Waterflood Installations," *JPT* (Feb. 1971) 173–79.
4. "Flow of Fluids Through Valves, Fittings, and Pipe," Technical Paper 410, Crane Co., New York City (1981).
5. *Piping Engineering*, Tube Turns Div., Louisville, KY (1979).
6. Mathur, A. *et al.*: "An Approach to Optimum Siting," paper SPE 10741 presented at the 1982 SPE California Regional Meeting, San Francisco, March 24–26.
7. Valaster, W.B., Burchard, H.G., and Sasscer, E.P.: "Nonlinear Least Squares Applied to Liquid Piping Design," paper SPE 14409 presented at the 1985 SPE Annual Technical Conference and Exhibition, Las Vegas, Sept. 22–25.
8. Campbell, E.E.: "Turnkey Waterflood Installation," *Proc.*, 8th Annual Southwestern Petroleum Short Course, Lubbock, TX (April 1961) 147–48.
9. Jowell, J.H. and Haberthur, C.R.: "Norge Marchand Deep Injection Project," *Pet. Eng.* (Nov. 1976) 23–33.
10. *Handbook of Steel Pipe*, SP-229-882-30M-SP, American Iron and Steel Inst. (1982).
11. Bartos, W.B.: "Two Calculator Programs for Hazen-Williams Pipeline Equation," *Oil & Gas J.* (April 30, 1984) 56–60.
12. Moore, B.L. Jr.: "Webster Field Unit Waterflood Facilities," paper SPE 6880 presented at the 1977 SPE Annual Technical Conference and Exhibition, Denver, Oct. 9–12.
13. EnDean, H.J.: "Materials and Installation Requirements for Handling Corrosive Waters," *Proc.*, 16th Annual Southwestern Petroleum Short Course, Lubbock, TX (April 1969) 211–17.
14. "Recommended Practice for Application of Cement Lining to Steel Tubular Goods, Handling, Installation and Joining," RP 10E, American Petroleum Inst., New York City.
15. Junkin, E.D. Jr.: "Corrosion and Corrosion Control of Oil and Gas Producing Equipment," *Proc.*, 13th Annual Southwestern Petroleum Short Course, Lubbock, TX (April 1966) 179–88.
16. Kohan, D.: "Criteria Given for Sizing and Selection of Control Valves," *Oil & Gas J.* (Dec. 16, 1985) 76–80.
17. Treadway, R.B., Roye, J.E. Jr., and Westerman, G.W.: "Solar-Powered Injection Controller Utilizing Bottomhole Pressure Sensing Device," paper SPE 9364 presented at the 1980 SPE Annual Technical Conference and Exhibition, Dallas, Sept. 21–24.
18. "Selection of Metallic Materials to be Used in All Phases of Water Handling for Injection into Oil Bearing Formations," Standard RP-04-75, Natl. Assn. of Corrosion Engineers, Katy, TX (April 1975).
19. Fowler, E.D. and Rhodes, A.F.: "Checklist Can Help Specify Proper Wellhead Material," *Oil & Gas J.* (Jan. 24, 1977) 62–65.
20. "Recommended Practice on Metallurgical Selection for Corrosion and Erosion Control," Arco Oil & Gas Co., Dallas, TX (June 1985).
21. Lubinski, A., Althouse, W.S., and Logan, J.L.: "Helical Buckling of Tubing Sealed in Packers," *JPT* (June 1962) 655–70; *Trans.*, AIME, **255**.
22. Hammerlindl, D.J.: "Movements, Forces and Stresses Associated with Combination Tubing Strings Sealed in Packers," *JPT* (Feb. 1977) 195–208; *Trans.*, AIME, **259**.
23. Sweet, R.G., Pumphrey, D.M. Jr., and Funk, R.J.: "Design Guidelines Help Select Fiber Glass Injection-Well Tubulars," *Oil & Gas J.* (April 6, 1981) 148–54.

SI Metric Conversion Factors

degrees × 1.745 329	E−02	= rad
°F (°F−32)/1.8		= °C
in. × 2.54*	E+00	= cm
lbm/ft × 1.488 164	E+00	= kg/m
psi × 6.894 757	E+00	= kPa

*Conversion factor is exact.

Chapter 7
Well Production and Injection Practices

7.1 Introduction

This chapter discusses the changes necessary in wells to upgrade and convert a field from primary production to a waterflood-response mode. Consideration of surface-facility consolidation and functional changes are covered in Chap. 8.

7.2 Injection Wells

Injection wells in a waterflood may be converted from producing wells or drilled specifically as injection wells. Such factors as waterflood pattern, well condition, primary-production-well pattern and spacing, and the stage of primary development will determine how many wells are drilled or converted. Chaps. 3 and 6 covered drilling of new injection wells. Conversion of existing production wells to injection requires a careful analysis of the available wells. Assuming that the existing well pattern and spacing allows the use of these wells as injectors, each well must be studied for economic suitability. In some cases, the cost of obtaining an adequate injection well by conversion may be more than drilling a new one. The condition of the existing well is usually the determining factor.

The temptation may be strong to convert marginal or low-capacity producers to water injection with the object of minimizing oil production loss; however, the poor productivity may be a result of low net pay or permeability values that would also cause low injectivity. It is frequently more economical to accept the higher initial loss of production to have a shorter response time for the waterflood.

7.2.1 Mechanical Condition.
The first step in evaluating wells for their suitability as injectors is to examine the mechanical condition of the wellbore. The engineer should check the pressure ratings, materials, and sizing of the wellhead. Generally, the casing head installed on a producing well will be adequate, but often the tubing head, master valves, and other wellhead parts will require replacement to meet future pressure requirements. Sec. 6.3.1 applies to the case where all or part of a producing wellhead is unsuitable for injection uses. The waterflood engineer will usually find it necessary to replace the downhole tubing and accessories with an assembly more appropriate for injection service (Sec. 6.3.2). A commonly used injection tubing, internally plastic-coated steel, is not normally used for producing wells. Hence, the tubing string is often changed. Packer and tubing assembly design should follow the principles outlined in Chap. 6.

The waterflood engineer should also check the condition of the well casing. If casing leaks or collapses are present within the field area, the wells should come under careful scrutiny. Casing inspection surveys may be cost-effective and appropriate to determine the extent of the problem. Casing and cementing records should be analyzed and calculations made as needed to ensure adequate strength and isolation for safe injection and compliance with regulations. Drilling engineers or consultants can assist with such calculations. It is of particular importance to ensure that the casing integrity will be maintained in the event of a tubing or packer failure that allows the injection pressure to be applied to the casing. Sec. 3.3.3 discusses other aspects of casing and cementing problems in more detail.

In some cases, producing wells will require redrilling, liner installations, or other mechanical improvements to be successful injectors. The waterflood engineer should consult personnel with drilling expertise for procedures and cost estimates to determine the most economical choice to condition a well for injection. In many cases, a combination of procedures may be best. In the Wilmington Townlot Unit,[1] 9 new wells were drilled for injection, 17 were redrilled, and 34 were converted. Some of the converted wells needed new liners installed. This waterflood is a good example of evaluating the desired injection wells individually to determine the most economical procedure.

7.2.2 Reservoir Condition.
A second general area the waterflood engineer must consider in converting producing wells to injection is the condition of the wellbore near the reservoir producing interval.

In some cases, the existing well perforations may not allow injection at the desired rates into the chosen intervals. This condition can frequently be remedied by reperforations, stimulations, cement squeezes, or other methods of controlling the injection profile. Fig. 7.1 shows a typical producing well that can present problems when being converted to injection service. For example, if it is desirable to inject only in Zone A (assuming the cement around the casing is good), then setting a bridge plug with cement on top or a cement plug over Zone B may achieve the goal. Conversely, if only Zone B should receive injection water, then setting a plug between zones and cement-squeezing the upper perforations may be the best solution. This would be followed by drilling out the plug to below the base of the desired injection interval with the injection packer set between Zones A and B. Setting the packer at that location without squeezing the upper perforations will generally be unsatisfactory because of probable violation of the casing-integrity requirements of the U.S. Underground Injection Control regulations (Chap. 10).

If several injection zones are planned in one wellbore, selective stimulations or reperforating may give the desired water-injection profile. Multiple tubing strings may also allow control of where the water is being injected but are expensive and frequently cause downhole operating problems. Ghauri[2] discussed the use of these techniques in the Wasson field of west Texas. Dual strings of tubing were found to be unsatisfactory, but several stimulation or plugging techniques (or both) were successful. Acid treatments were used, and old perforations were found to need about 1.5 times as much acid volume per foot as new perforations to achieve good results. In some cases, fracture treatments have been acceptable stimulation techniques but they require very careful control.[3] The risk of fracturing out of the zone or bypassing oil near the injection wellbore must be balanced against the better sweep performance achieved by a successful stimulation.

One problem sometimes found when converting a producer is poor injectivity resulting from relative permeability effects.[4] Relatively high oil saturations can interfere with the flow of water through the zone. Removing the oil can increase the injectivity. Some improvement will be made as the water reduces the oil to residual saturation levels during the first few days of injection. Further reduction in oil saturation can be accomplished by a variety of methods,[5,6] including the use of micellar solutions, mutual solvents, and surfactants. In the Magnus field, Dymond and Spurr[6] found that surfactant applications increased injectivity by about 15%, but fracture treatments were more economical stimulations for the widely spaced wells.

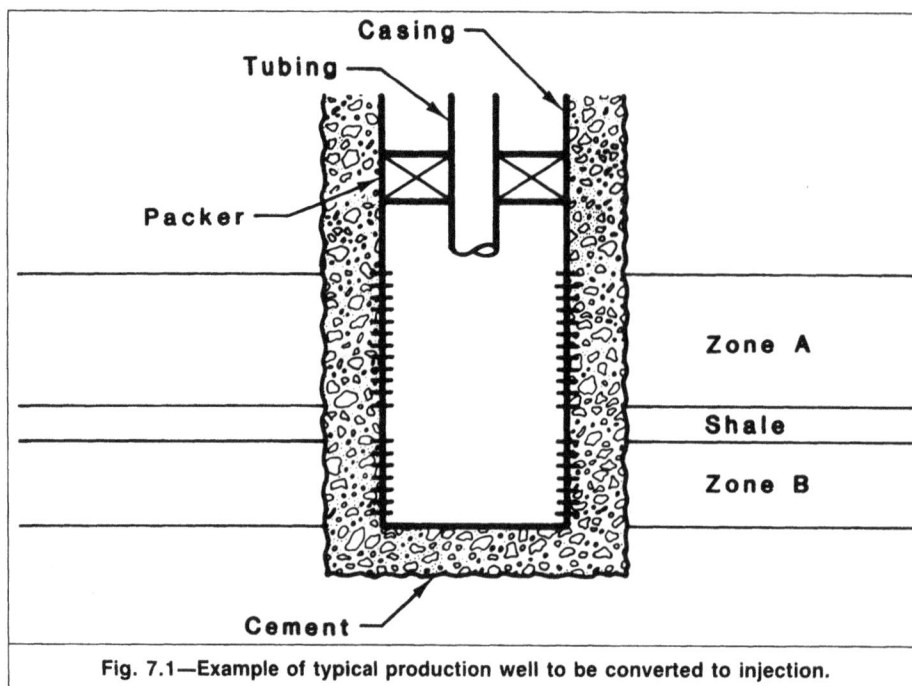

Fig. 7.1—Example of typical production well to be converted to injection.

A brief overview of some problems and solutions in converting a producing well to injection has been presented. The waterflood engineer should consult drilling or engineering personnel with expertise in completion and stimulation design for complex cases. Various services companies can also offer assistance.

7.3 Producing Wells

The equipment used for primary production is very likely inadequate to handle the significantly larger volumes of water produced at the maximum waterflood-response level. Depending on the timing of waterflood, the equipment on hand may be suitable during the early periods. Particularly if the flood is begun relatively soon in the primary life of the field, the artificial-lift and treating facilities may be able to handle the production at least for some time before full response occurs. The waterflood engineer should use predictive techniques, such as inflow performance relationship curves and nodal analysis, to determine the expected rate of production per well.[7] This information can then be used to select the type of artificial lift and to size the equipment properly.

7.3.1 Artificial Lift. As a general rule, producing wells should be kept pumped down as much as possible in a waterflood to minimize the backpressure on the reservoir sandface. One rule of thumb is to keep the downhole producing pressure at or below 10% of the average reservoir pressure in the area. The reservoir engineer should supply information by which the required or expected rates can be estimated on a per-well basis with the techniques mentioned above. Assuming that the predicted rates are greater than the existing equipment can lift, the waterflood engineer must plan on installing larger units or different types of lift. The actual installation of the new equipment may be phased in over time as the waterflood progresses. In some cases, larger tubing strings may be required to accommodate higher production rates.

A full discussion of artificial-lift equipment is beyond the scope of this monograph, but a few guidelines follow. For small to medium volumes (up to a few hundred barrels per day) from < 10,000-ft [3048-m] depths, rod-pumping systems are usually the most economical to install. At shallower depths, larger volumes can be pumped. For large volumes, gas-lifting and electric submersible centrifugal pumps may be suitable for up to 10,000 B/D [1590 m^3/d] or more. The exact rates and depths available vary with the

equipment chosen, temperatures, and other bottomhole conditions. Hydraulic pumps, either piston or jet types, fall in the intermediate-volume ranges, with depth capabilities of > 10,000 ft [3048 m]. The economics of the "best" system will depend on many factors, such as cost and availability of fuel or power, local servicing capability, and well-fluid characteristics. Ref. 7 is a good source for further information on artificial lift.

Various types of controllers are available for different types of artificial lift, particularly for the positive-displacement and centrifugal pumps. Various forms of pumpoff controllers allow wells to be kept pumped down as much as possible while eliminating or minimizing damage to pumps. For example, pumpoff controllers were successful in improving sucker-rod pumping in the McElroy field.[8]

The state of the art in artificial-lift technology is constantly changing. The information given here provides only rough guidelines. The waterflood engineer should consult the literature and knowledgeable persons for the latest information with which to develop the waterflood design.

References

1. Nelson, K.C.: "Wilmington Townlot Unit-Waterflood Implementation and Response," paper SPE 6167 presented at the 1976 SPE Annual Technical Conference and Exhibition, New Orleans, Oct. 3–6.
2. Ghauri, W.K.: "Production Technology Experience in a Large Carbonate Waterflood, Denver Unit, Wasson San Andres Field," *JPT* (Sept. 1980) 1493–1502.
3. Tinker, G.E.: "Design and Operating Factors That Affect Waterflood Performance in Michigan," *JPT* (Oct. 1983) 1884–92.
4. Hensel, W.M. Jr., Sullivan, R.L., and Stallings, R.H.: "Understanding and Solving Injection Well Problems," *Pet. Eng.* (May 1981) 155–70.
5. Moran, J.P. and Maxwell, W.N.: "Field Results of Injection Well Stimulation Treatments Using Micellar Dispersions," paper SPE 2842 presented at the 1970 SPE Practical Aspects of Improved Recovery Techniques Symposium, Fort Worth, March 8–10.
6. Dymond, P.F. and Spurr, P.R.: "Magnus Field: Surfactant Stimulation of Water-Injection Wells," *SPERE* (Feb. 1988) 165–74.
7. Brown, K.E.: *The Technology of Artificial Lift Methods*, PennWell Publishing Co., Tulsa, OK (1984) 1–4.
8. Amezcua, J.D.: "Pump-Off Controllers Improve Sucker Rod Lift Economics," *World Oil* (Feb. 1, 1982) 55–60.

Chapter 8
System Consolidation and Automation

8.1 Introduction

Conversion of a field from primary operations by one or more operators to a waterflood operated by a single party frequently offers an opportunity to consolidate facilities for more cost-effective operation. Consolidation will generally be economically feasible, especially if the field or unit contains numerous small facilities. Automation of operations during the process of converting to waterflooding may also prove to be a good investment. Decisions to consolidate or to automate must be justified economically on a "stand-alone" basis. The costs of such work often have been included in the waterflood investment and justified by the waterflood cash flow. This is an incorrect approach. The engineer should evaluate the merits of consolidating or automating independently from waterflood economics.

8.2 Production Facilities Consolidation

In most cases, the engineer will be designing a waterflood for a mature field producing under a primary depletion mechanism. Frequently, several independent operators will be producing different parts of the field, resulting in a number of small production facilities. Assuming that the waterflood is a unitized or cooperative (see Chap. 11) operation, an opportunity exists to optimize the number of facilities for more efficient operation. At the Wilmington Townlot Unit in California,[1] the engineers designed a system that centralized 360 batteries and more than 300 leases into only 33 satellite stations and one central treating facility. Although it was not part of a waterflood installation, an operator[2] consolidated some 913 wells and 93 lease facilities in the East Texas field into a single facility and saved significant amounts of lease operating costs.

8.2.1 Existing Facilities. The waterflood engineer should survey the existing field facilities to evaluate potential savings from relocation or abandonment. The survey could include consideration of (1) replacement of worn, inefficient, and high-cost equipment; (2) replacement of inadequately sized equipment; (3) relocation of facilities to achieve operating efficiencies that may include less manpower required, less maintenance of roads and lines, and better distribution of fuel or electricity; (4) safer working conditions; (5) less environmental damage; (6) better match to the waterflood facility; and (7) simpler produced-water collection and handling.

8.2.2 Consolidation. The replacement of worn, inefficient, and high-operating-cost equipment is generally associated with replacing equipment of inadequate size. A waterflood normally produces larger liquid volumes than primary depletion. Hence, the available equipment is frequently too small for the expected waterflood volumes. Production facilities can be centralized by moving numerous small production vessels that are available to a central location. However, this procedure usually results in a costly and inefficient operation that requires substantial surveillance and maintenance. At the North Coles Levee field in California, an attempt was made to use existing small vertical separators and equally small manifolds and tanks to handle waterflood production in 8 to 10 batteries. The costs of repairs and maintenance were quickly found to be excessive, as was the cost of getting produced water to the

central water treatment station. Three flow stations were installed with new, automated manifold systems and large-diameter, horizontal separators. A new water gathering system from these flow stations was also installed. The resultant savings were more than adequate to justify the investment. Flowline sizing should also be evaluated by the engineer. In the example cited above, replacing 2- to 3-in. [5.1- to 7.6-cm] lines with 6- to 8-in. [15.2- to 20.3-cm] lines was found to be completely justifiable.

Relocation of facilities must be examined during the design of the waterflood. Minimizing the number of facilities and locating them on a well-planned road and utility system can reduce operating costs substantially. Fewer personnel will be required for operating and maintenance, and centralized utility metering may reduce expenses. Relocation can also help to match the production to the water-injection facilities. As discussed in Chap. 6, injection plants and lines should be designed to minimize the initial investment and operating costs. Centralizing the production facilities at the same locations as the waterflood plants is frequently cost-effective. This arrangement also simplifies produced-water collection and minimizes the handling of such water.

As mentioned in the discussion of water-injection distribution in Chap. 6, it may be most economical to gather the production in satellite stations and then transfer the fluids to a central processing battery in a large "group" line. Testing can be performed either at the satellite stations or, through a test line, at the central battery with a test separator. This layout is similar to the trunk and radial systems for water injection discussed in Chap. 6.

In surveying existing facilities, the engineer must also consider personnel safety, equipment, and the environment. The introduction of high-volume water processing and high-pressure water injection will likely produce more hazardous conditions. Any deficiencies in existing facilities must be remedied in the waterflood design. The costs of retrofitting equipment to provide adequate protection lend further economic justification to consolidation and to upgrading proposals.

While the waterflood engineer should carefully consider using all on-hand equipment that is suitable, older fields will usually justify a substantial amount of consolidation, replacement, relocation, and upgrading of the production systems to process the waterflood volumes effectively. In such fields as Kuparuk on the North Slope of Alaska, where the original development design included the waterflood facility, this effort would not be required. The most common case that occurs, however, is unitization of an older field for waterflooding, so the waterflood engineer will generally need to evaluate production facility consolidation.

8.3 Automation

Automation of production facilities is covered in a number of technical papers.[3-6] This constantly changing area of production engineering could justify a volume by itself. This monograph does not discuss this field of technology in depth. The engineer is advised to consult the literature and experts on automation of production facilities. Automation of water-injection and treatment facilities is covered in Sec. 8.3.2.

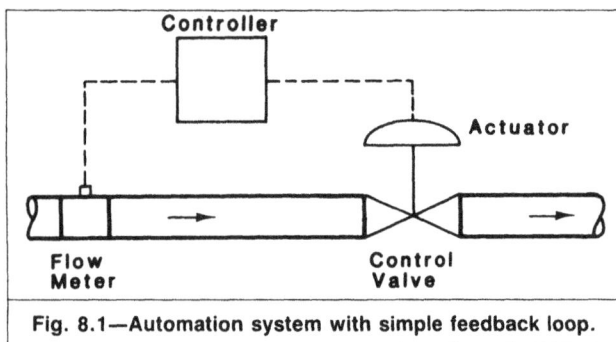

Fig. 8.1—Automation system with simple feedback loop.

8.3.1 Concepts. In his excellent paper, Harrison[4] defines automation as "a machine system that through measurement, control and feedback will allow operation with optimum efficiency." One key element in this definition is feedback. Without a feedback loop that drives the control function exercised, the system will not operate efficiently under changing conditions. The heart of any automation system is an "intelligent" machine that receives information from the feedback mechanism and directs action. These devices are generally computers that range from small microprocessors to large, "main-frame-type" computers. The software programs also range from very simple to very complex. The major differences are the number of inputs/outputs and the difficulty of the decisions (actions) that must be made.

If a computer is the heart or brain of an automation system, then the system's eyes, ears, and fingers are the "end devices." There are two major categories: sensing and control devices. Examples of the sensors include (1) temperature-sensing instruments (thermometers and thermocouples), (2) pressure-sensing instruments (Bourdon tubes and transducers), (3) flow-rate-measuring devices (turbine, orifice, and vortex meters), (4) composition or quality measurement instruments (turbidity meters, moisture analyzers, and H_2S-sensing devices), and (5) condition-sensing instruments (level-measurement floats, valve position monitors, flow/no-flow sensors, and on/off power switches). This list is not all-inclusive but gives some of the more common sensors found in oilfield operations. Virtually anything that can be measured can be included in an automation system.

The second category of end devices is the finger or hand of the systems, the mechanisms through which the control function is exerted. A few examples would include engine-speed controllers, switches that turn equipment on or off, control valves, chokes, and variable frequency controllers.

The end devices that collect data, the computers that analyze the information and issue commands, and the control devices that perform actions can be combined to achieve automatic control for almost any type of process. Fig. 8.1 shows a simple system consisting of a microprocessor that receives a signal from a flow-rate meter and sends a signal to a control-valve actuator, which throttles the valve to set a new flow rate. The meter is then used to check that the correct rate is flowing. More complex systems can have intermediate levels of control and measurements. Some authors believe that each level should be able to function on its own. Morgen and Dozier[7] express the philosophy that "each unit within an overall system must be able to independently control its own process." If the definition of a "unit" is broad enough, this is probably a good guideline. Current design principles would not apply this concept to each end device and control function. But if a "unit" is considered to be a complete process taking place in a compact piece of equipment, then current practice would meet their philosophy.

Fig. 8.2 shows a more complex automation scheme that includes a satellite well manifold and test system feeding a central treating facility. The satellite remote terminal unit (RTU) controls the switching of wells from group to test lines, records the results of the tests, and transmits that information to the central control unit. Although not shown, the RTU could also collect, transmit, or take action on other information such as pressures, temperatures, pumpoff status of wells,[8,9] or flow/no-flow condition. The data flow from the RTU to the central unit and back can be over a wire[5] or optic fiber[6] system ("hard-wired") or sent with a radio or microwave transmitter. Fig. 8.2 shows a central treating facility also operated by the central control unit. As an example of the functions that can be included in automation systems, this particular case shows the central unit monitoring the level in a water tank and controlling the flow into the tank by switching valves on the inlet lines, thus shifting the water from the tank to discharge or the reverse. In light of the philosophy ascribed to by Morgen and Dozier,[7] it would be a better design if the level-measurement device would directly control the valves in the absence of a signal from the central unit.

While Fig. 8.2 shows a more complex system than Fig. 8.1, even more complicated automation is common. At Upper Zakum field,[6] a supervisory control and data acquisition (SCADA) system allows the operator to handle up to 500 wells, 100 platforms, and more than 1,000,000 BOPD [159,000 m³/d oil]. This system has 74 RTU's transmitting to four main terminal units, which, in turn, are in communication with a central control room. Some 36,000 end devices are handled by this SCADA system. Other examples of very complex systems can be found in Refs. 2, 3, 5, and 6, including the Prudhoe Bay and Wasson San Andres fields.

Harrison[4] states that it "is more efficient, faster and easier to train expert production operations personnel in computer and communications technology than to train computer and communications experts in production technology." While this may be correct for

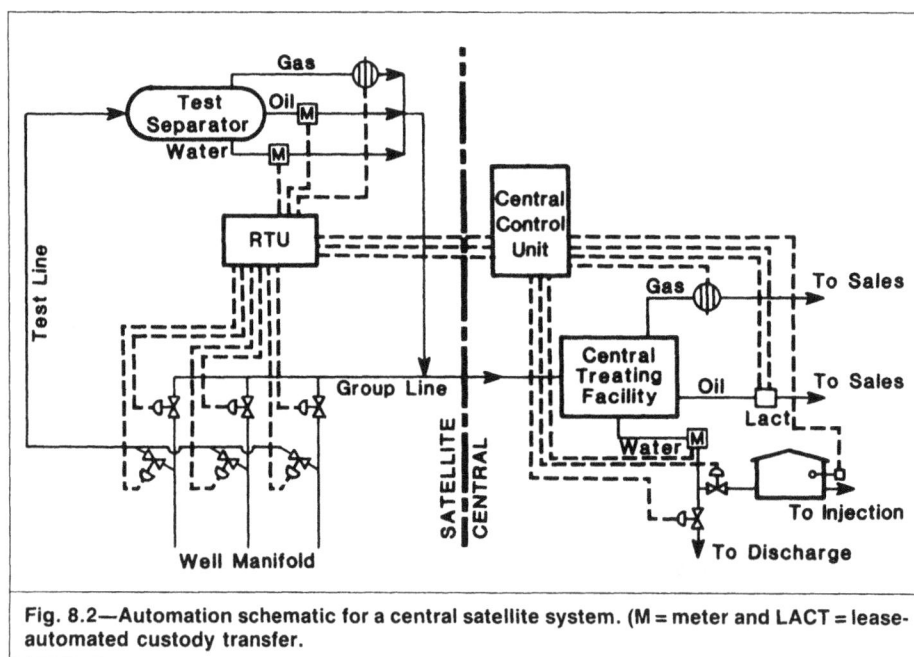

Fig. 8.2—Automation schematic for a central satellite system. (M = meter and LACT = lease-automated custody transfer.

normal operations, the maintenance, repair, and reprogramming of such systems should be done by specialized personnel with the appropriate technical training. In the Webster Field Unit,[10] the designers decided to use standard oilfield instruments and controllers for the end devices in the automation system. The technique allowed the maximum use of the talents and expertise of the personnel already on site to operate, maintain, and repair the system. Personnel training covered the operation of the system. The engineer must include the costs of training and of specialized contract labor for maintenance and repair in the economic evaluation of the automated system.

As mentioned previously, the waterflood engineer involved in automation system design for production (or injection) facilities should refer to the current literature on the subject, vendor companies selling such equipment, and in-house or outside consultants expert in such systems. Any automation that is contemplated must be economically justified on its own merits separately from waterflood economics.

8.3.2 Injection Systems. Three separate areas for potential automation in waterflood injection systems are (1) the water gathering systems, treatment equipment, and wells (source and produced waters); (2) water-injection plants; and (3) water-injection wells. Control can be exercised at various levels within these areas. All three areas can be automated and controlled from one point with a centralized automation system. Alternatively, each system or even each piece of equipment and well can operate independently under its own automated process control. Each waterflood has individual characteristics that determine the degree and type of automation that optimizes the economic operation of the waterflood. At Prudhoe Bay, for example, the source and treating systems are fully automated on an independent basis. Each injection plant and associated wells are operated by a common control. However, each source, treatment, injection plant, or injection-well system can function independently or on local (in some cases, manual) control, if necessary. At the Hewitt Unit[11] in Oklahoma, the engineers designed a system in which the source-water wells, booster pump plant, and injection plant are connected and balance flow rates to keep the total system pressures and rates within limits. For example, if the suction line pressure to the injection pumps decreases, the booster pump speed is increased to supply more water to the injection plant, which is 8 miles [13 km] away. However, no central control from a single point is exerted over the entire process. Again, it must be emphasized that each system must be designed to fit the conditions of the specific project.

8.3.2.1 Source and Treating Systems. Source wells are frequently controlled using on/off or speed controllers, depending on the type of artificial lift (if any) being used in the wells. At the Hewitt Unit,[11] four source wells with submersible pumps were turned on or off as needed, while a speed-controlled, engine-driven, shaft turbine pump was used in one well to handle routine fluctuations in demand. As a general rule, this type of concept—i.e., supplying a base load with wells operated on an on/off basis and handling normal variation with a variable rate system—is a valid design principle for source or other pumping processes. The information upon which the control function is based may be the booster or injection pump demand rate as measured by pressure or tank levels. Ultimately, source rates are controlled by the injection-well rates and limited by the source-well capacities. Any automatic control system must be designed with this basic idea in mind.

Water treating equipment is generally supplied with individual automation systems. They may vary from mechanized systems for cartridge filters and flotation cells to complex computer-based systems for deaeration towers or media filters. In most waterfloods, these control systems prove adequate and no further automation is required. In such cases, monitoring and transmitting instrumentation should be evaluated to give the operator current status information. For cases in which further control is needed, the engineer should consult companies that supply the equipment for the best way to fulfill this need and link the two systems.

8.3.2.2 Water-Injection Plants. Automation systems in water-injection plants will normally be designed for three major purposes: (1) to supply the required rates to the injection wells at the neces-

sary pressure; (2) to prevent damage to equipment and personnel; and (3) to aquire and store data. Flow-rate control can be achieved by regulating the number and speeds of the injection pumps, by bypassing water back to the suction tank or system, or on centrifugal pumps, by varying the discharge pressure by controlling the pressure drop across an orifice or valve on the downstream side. Combinations of these techniques may also be used.

Regulating the number and speed of the pumps generally costs more for the initial investment in the system. If multiple smaller pumps are used, maintenance and operation costs are larger. The other two control techniques involve increased or wasted power and fuel usage, thus increasing operating costs. These procedures require the installation of more powerful pumps and prime movers than are needed to supply the pressures and rates desired at the injection wellheads. This increases initial cost and decreases overall efficiency because of the wasted energy. The waterflood engineer must make an economic comparison based on the choices available in the waterflood design and determine the best system(s) to install.

In many cases, separate and different automatic controls have been used on the produced- and the source-water injection systems.[10,11] The produced water must generally be handled properly to maintain the field oil rate. In the short run, source water may be reduced without affecting oil production. Most automatic systems will therefore provide that the produced-water controls will override the source-water system. Some systems are designed so that additional pumps or pumping capacity will switch to produced-water service as needed. For example, a level control on the produced-water tank can be used to vary the source-water flow to the suction tank of the injection pumps. If the level increases, the source-water rate is reduced, giving more pump capacity to the produced water and, hence, reducing the tank level.

Regardless of the technique used, most injection-plant controls use information on either pressure or rate to determine how the control functions. At the Hewitt Unit,[11] the suction or the discharge-manifold pressures control the engine speed for the source-water injection pumps. The produced-water pumps are speed-controlled to the metered discharge rate that can be overridden by a produced-water tank level signal. Another very common control system is a governor or speed controller on the pump engine that is regulated by a suction tank level instrument. As a guideline, the engineer should select the least costly and simplest system of automation that achieves the design function.

8.3.2.3 Water-Injection Wells. In most waterfloods, the parameter that finally controls all rates and pressures is the sum of the injection wells' rates and the pressures required to achieve these rates. Most injection wells have some type of control device but automation is frequently unnecessary. Chap. 6 outlines some methods used for wellhead control. These devices may be as simple as a positive choke that forces the flow through a fixed orifice, thus restricting the rate. Such a control mechanism requires manual changing of the set point by changing the choke size, and hence, is not suitable for automation. The control equipment that will be described below can be automated. As discussed earlier, however, automation does not mean that each well must be connected to and controlled by a centralized, single- or multiple-point computer system.

The most fundamental controls that are exerted on an injection well are flow-rate and pressure regulation. Other functions, such as alarm and shutdown, may be necessary, but rate and pressure control are the most important for normal operation of the well. The two most common types of control devices are the control valve and the choke. In automated systems, this equipment will have some method of directed operation, usually actuators driven by hydraulic, pneumatic, or electrical forces. For flow-rate control, meters will frequently supply the information for the feedback loop to the controller that directs the movement of the valve or choke. For pressure control, a gauge, pressure regulator, or transducer will perform that function. Chokes are simple devices that have an orifice or hole through which the flow passes. Changing the size of the hole allows more or less flow to pass with the same pressure drop. An automated system can use a choke in which the orifice size is adjustable. The manufacturers of chokes can supply the engineer with the necessary data on choke coefficients, sizes, materials, and

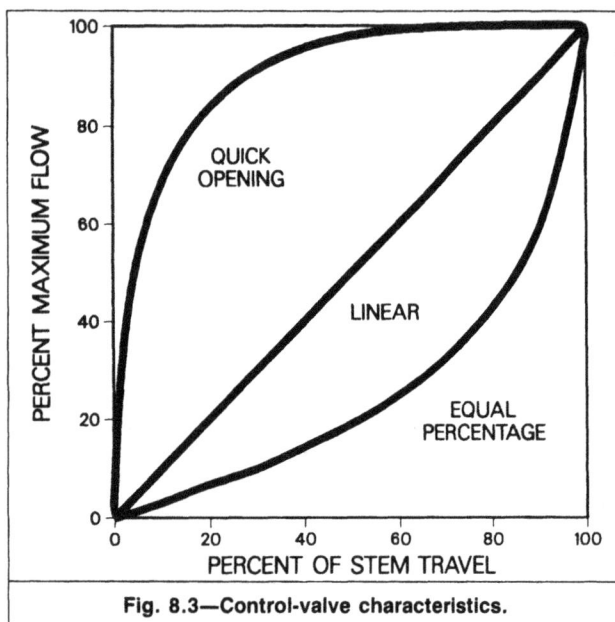

Fig. 8.3—Control-valve characteristics.

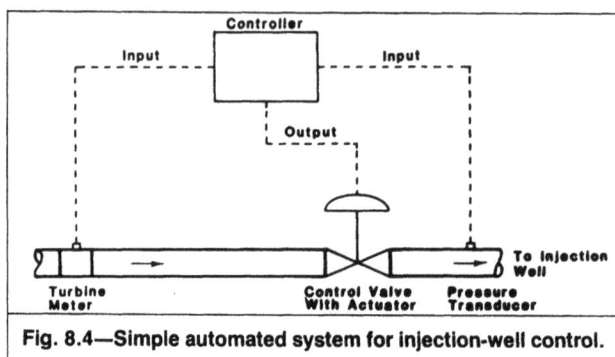

Fig. 8.4—Simple automated system for injection-well control.

actuators. Chokes are normally subject to high pressure drops, cavitation, erosion, and corrosion; therefore, material selection is particularly important. Vendors, trade and scientific literature, and consultants can assist the engineer in the selection of materials for chokes and control valves.

Control valves perform a function similar to adjustable chokes. Both create a variable orifice for the fluid to move through, changing pressure and rate in the process. Valves and chokes can be located at the wellheads or at a manifold. The latter positioning makes automation simpler, but necessitates individual lines to each well from the manifold; i.e., a radial system rather than a trunk and lateral system will be required downstream of the manifold. Hence, this design may increase the total length of the injection lines. The engineer should evaluate both cases to ensure that the most cost-effective choice is made. There are three different characteristics of control valves, as shown in Fig. 8.3. Other curves can be obtained to fit a particular application by special designs and nonlinear positioners on the valve actuators. Generally, ball- and butterfly-valve characteristics are equal percentage. Globe and angle valves are available with quick-opening, linear, or equal-percentage characteristics. Normally, gate and plug valves are not used for throttling control service, although certain specialized versions of these valves do have applications. More rarely, such valves as pinch or diaphragm types are used. In waterflood service, gate, plug, and ball valves are most commonly used as block valves with globe, angle, and ball valves used for throttling (control) purposes. An excellent report showing diagrams and discussing applications can be found in Ref. 12. Numerous other articles have been written on types, sizing, and materials of control valves used in the petroleum industry.[13-15] The waterflood engineer should consult the literature and other sources of expertise when designing the flow-control systems.

The combination of a final control element, such as a valve or choke, with an actuator is the system that supports automation. Actuators are available to drive with either linear or rotary motion. The driving force can be electric, hydraulic, or pneumatic. The device being driven can be mechanical or hydraulic. The fundamental parameters the engineer must consider are the type of motion required for the control element and the torque or power needed for full travel over the desired range. Manufacturers and the literature will provide the engineer with the specifications of the choke or valve and the available actuators to allow the proper match to be designed. Positioners monitor the actual valve travel and continue actuator movement until the desired adjustment is reached. These instruments should be considered because they ensure accurate and often faster control response.

Another essential part of an automated control system for injection wells is the instrumentation that provides the data for the feedback loop. The most common type is the meter, although pressure

may be the controlling factor, requiring that transducers, gauges, or regulators be used. Because rate is normally the control parameter, meters will be discussed below. Pressure-regulating devices are usually in place as an override type of control; i.e., rate controls unless the upper or lower pressure limits are reached, at which time control diverts to the pressure devices.

Flowmeters are of various design types, including positive displacement (PD), orifice, turbine, and vortex. The first three are the most common in waterflood operations. PD meters are the least expensive but are not suitable for automation without expensive accessory equipment. Orifice and turbine meters frequently are similar in expense when equipped for automation. However, turbine meters have a wider range of measurement for a given size and are easier to maintain. The use of turbine meters in waterflood service is recommended in most cases, particularly if automation is contemplated.

While extreme accuracy normally is not needed in individual well meters, turbine meters give accuracies to 0.1% with a similar value for repeatability. Meters of any type should have a specific program of calibration and maintenance to ensure the desired accuracy level. Master meters, whose measurements can be compared with the sum of other meters' readings, may be helpful in monitoring accuracy. Pressure taps on either side of a meter to measure the pressure drop across the instrument help detect problems requiring repair or replacement. Meter provers that allow direct testing of a meter's accuracy and repeatability may also be justified. In any case, a program of inspection, monitoring, calibration, and maintenance is necessary to ensure a properly functioning metering system.

The different parts of an injection-well control system, as described above, can be assembled into an integrated unit that provides proper control. Westerman[16] describes one such system that includes a ball valve, electromechanical actuator, turbine meter, strain-gauge transducer, and microprocessor (Fig. 8.4). The system can be set to control rate or pressure or both on an override condition. Another stand-alone automation system was developed for the North Cowden Unit in west Texas.[17] In this case, a bottomhole-pressure sensor supplied the signal to the microprocessor, which adjusted the position of a ball valve. The purpose was to maximize the water-injection rate without exceeding the fracture pressure of the formation. Again, in this case, turbine meters and pressure transducers were used for data acquisition.

Independent, integrated systems involving data sensors, controllers, and control elements, as described above, constitute the most common form of automation in waterfloods. More complex systems normally involve many such independent systems that enter data into a centralized computer that then controls the set points at the well or manifold. The telemetry of data and commands may involve transmission by wire, optic fiber, radio, and microwave to and from the central point. Different levels of control may be involved, such as at Prudhoe Bay,[5] with independent control possible at the well pads and/or control possible with the SCADA system from the gathering process center or by the central operator at the main operations center. Such central systems frequently include controls for wells, plants, and source systems. The fundamentals described in this section are applicable to large and complex or small and simple systems. As mentioned before, the waterflood engineer faced with designing an automation system should refer to sources of information, including current literature, in-house or outside consultants, and manufacturers. A good guideline to designing these systems is a quote from Albert Einstein: "Everything should be made as simple as possible but not simpler."

References

1. Nelson, K.C.: "Wilmington Townlot Unit—Waterflood Implementation and Response," paper SPE 6167 presented at the 1976 SPE Annual Technical Conference and Exhibition, New Orleans, Oct. 3–6.
2. Abel, W.L. and Pace, F.A.: "A Lease Consolidation, Telemetering, and Supervisory Control Project," JPT (Jan. 1968) 12–16.
3. Ghauri, W.K.: "Production Technology Experience in a Large Carbonate Waterflood, Denver Unit, Wasson San Andres Field," JPT (Sept. 1980) 1493–1502.
4. Harrison, O.R.: "Planning and Implementing a Computer Production Control System," JPT (Feb. 1970) 134–38.
5. Hardy, I.J. and Wetzel, G.P.: "Automated Production Systems," paper SPE 10005 presented at the 1982 SPE Intl. Petroleum Exhibition and Technical Symposium, Beijing, March 18–26.
6. "New SCADA Technology Used To Operate Upper Zakum Field," Pipe Line Ind. (Oct. 1984) 26–27.
7. Morgen, R.A. and Dozier, H.S.: "A Stepwise Approach to Lease Automation," paper SPE 353 presented at the 1962 SPE Upper Gulf Coast Drilling and Production Practices Conference, Beaumont, TX, April 5–6.
8. Amezcua, J.D.: "Pump-off Controllers Improve Sucker Rod Lift Economics," World Oil (Feb. 1, 1982) 55–60.
9. Lea, J.F.: "New Pump-off Controls Improve Performance," Pet. Eng. (Dec. 1986) 41–44.
10. Moore, B.L. Jr.: "Webster Field Unit Waterflood Facilities," paper SPE 6880 presented at the 1977 SPE Annual Technical Conference and Exhibition, Denver, Oct. 9–12.
11. Cox, E.: "Surface and Reservoir Control of Injection Water in the Hewitt Unit Waterflood," JPT (April 1972) 426–30.
12. O'Keefe, W.: "A Special Report: Control Valves, Actuators, Regulators, Positioners," Power (April 1976) S-1–S-16.
13. Kohan, D.: "Criteria Given for Sizing and Selection of Control Valves," Oil & Gas J. (Dec. 16, 1985) 76–80.
14. Schumacher, M. and Gossett, J.: "Selecting Materials for Recirculation Valves in Secondary Recovery Service," Oil & Gas J. (Feb. 28, 1983) 98–102.
15. Bertrem, B.: "Butterfly Valves in the Petroleum Industry," Proc., 22nd Annual Southwestern Petroleum Short Course, Lubbock, TX (1975).
16. Westerman, G.W.: "Improved Controls for Enhanced Recovery," Proc., 29th Annual Southwestern Petroleum Short Course, Lubbock, TX (1982) 339–48.
17. Treadway, R.B., Roye, J.E. Jr., and Westerman, G.W.: "Solar Powered Injection Controller Utilizing Bottomhole Pressure Sensing Device," paper SPE 9364 presented at the 1980 SPE Annual Technical Conference and Exhibition, Dallas, Sept. 21–24.

Chapter 9
Economic Evaluation

9.1 Introduction

This chapter discusses the economic evaluation of a waterflood as a whole, but similar analyses are necessary at many stages of waterflood design, as noted in previous chapters. The principles are the same regardless of the type and scope of the specific problem involved. This chapter addresses only the economic calculations before federal (or state) income tax (BFIT) is considered. Each company will have a different tax calculation procedure. The waterflood engineer should use the appropriate method to derive after federal income tax (AFIT) values.

9.2 System Cost Estimation

Two primary cost types must be included in the economic analysis of waterfloods: investment (capital) and operating (expense) costs. These costs are treated differently in the evaluation and must be kept separate, especially for AFIT calculations.

9.2.1 Investment. The major costs associated with initiating a waterflood are capital investments. These costs generally fall into several categories, such as tangible and intangible drilling costs, workover/conversion of wells to injection, source/gathering/injection facilities, and consolidation or upgrading of production facilities. The earlier chapters covered the design of the waterflood facilities that must be in the cost estimate. Whether capital or expense, all costs should be included. The timing of the investments also needs to be estimated to develop a cash flow tabulation for the evaluation. Appendix C defines the term "cash flow" and other economic parameters used in this chapter. A typical list of the expenditures that may be required in a waterflood are shown in Table 9.1.

Costs should be determined with reasonable accuracy. Some sources for estimates include recent quotations for equipment and installation, manufacturers and their representatives, in-house purchasing agents, local construction firms, and engineering and operating personnel. Most manufacturers and vendors are willing to supply an initial, "engineering-quality" estimate for their equipment. This value is not a detailed quotation and will tend to be higher than an actual bid to supply the item. However, the accuracy will usually be adequate for the purpose of economic evaluations. Requesting such an engineering estimate is recommended when recent quotations are not available for major items of equipment. Labor costs are usually best obtained from local company production or construction supervisors. For major projects, construction companies may be contacted for information, but such estimates must be used cautiously because of the wide variation possible between companies.

Other costs that must be included are sales tax and transportation (freight) charges on items of equipment to be purchased. Vendors can usually supply this information or company purchasing departments may be able to help. It is also suggested that the engineer include some contingency funds. For major units such as turbines, pumps, engines, treating systems, and production vessels, a factor of 5 to 7% is generally adequate. A contingency allowance of 15 to 25% of the cost of the remaining expenditures after subtracting the major items is recommended. The engineer should base the amount of the contingency factor on the accuracy and completeness of the design and cost estimate. An example of a cost estimate (dated late 1985) for a Wyoming waterflood is shown in Table 9.2. In this case, the contingency funds include an allowance for unforeseen risks in the well work and production facilities because the condition of some wells and tank batteries was not known by the estimator. The total contingency amount was about 25% of the estimate. Future investments required can be calculated similarly, with inflation included at some estimated rate. Most companies have an inflation forecast available for consultation by the engineer.

9.2.2 Operating Costs. The engineer must also include the cost of operating the waterflood in the economic evaluation. The important factor is the incremental cost of the waterflood compared with continued primary production. Operating costs for the properties before waterflooding can be averaged to get a reasonable estimate for the cost per barrel of fluid produced under the primary phase. These costs, multiplied by the remaining primary production, can be subtracted from the estimated waterflood operating expenses to get incremental values.

TABLE 9.1—TYPICAL WATERFLOOD EXPENDITURES

Source or Supply
  Wells—drill and equip
  Surface—intakes and equipment

Gathering and Treatment
  Lines, manifolds, controls, and pumps
  Treating—skimmers, deaerators, and filters

Injection
  Booster/injection pumps and prime movers
  Controls, instrumentation, electrical system, and fuel
    systems
  Suction and discharge piping
  Foundations, buildings, tanks, and safety equipment

Distribution
  Lines, manifolds, controls, and instrumentation

Injection Wells
  Drill and equip
  Convert existing wells and equip
  Profile-control work

Production Wells
  Drill and equip
  Workovers
  Profile-control work
  Add or upgrade artificial-lift equipment

Production Facilities
  Consolidation and upgrading of process systems,
    including gathering lines, liquid and gas treating
    facilities, and test equipment
  Automation and telemetry systems

TABLE 9.2—EXAMPLE OF A WATERFLOOD COST ESTIMATE (1985 dollars)

Field is located in the Powder River basin, northeast Wyoming. The reservoir is an aeolian sandstone at an average depth of 8,700 ft (2650 m). A peripheral linedrive was chosen as the flood pattern.

Water Supply—Convert Well No. 3 to source well			
Well workover		$ 32,000	
New tubing string		11,500	
Electric submersible pump		60,000	
Miscellaneous		15,000	
		$118,500	
	Contingencies	15,500	
	Subtotal		$134,000
Injection Facility			
Injection pumps, motors, controls, and associated piping and electrical equipment in insulated building		$160,000	
Source-water tank		13,500	
Produced-water tank		17,500	
Location preparation		10,000	
Electric power system		14,000	
Piping		12,000	
Trucking		2,500	
Miscellaneous		10,000	
		$239,500	
	Contingencies	35,500	
	Subtotal		$275,000
Injection Distribution System			
Plastic-coated steel line pipe		$ 65,000	
Laying costs and labor		27,500	
Trucking		3,000	
Damage claims by landowner		7,000	
Miscellaneous		13,000	
		$115,000	
	Contingencies	17,000	
	Subtotal		$132,500
Injection Wells—Convert Wells No. 1, 2, 4, and A-4 to Injection			
Well workover costs		$ 93,000	
New tubing		162,000	
Wellheads		15,000	
Miscellaneous		21,000	
		$291,000	
	Contingencies	98,500	
	Subtotal		$389,500
Revision of Production System			
Consolidate batteries with new lines		$ 54,700	
New treater and free water knock-out		47,000	
New insulated building and location		25,000	
Damage claims		4,200	
Trucking		3,000	
Miscellaneous		12,500	
		$146,400	
	Contingencies	63,600	
	Subtotal		$210,000
	Grand Total		$1,141,000

TABLE 9.3—ECONOMIC EVALUATION PRESENTATION, INCREMENTAL CASH FLOW (BFIT) WITH INFLATION

Parameter	Year 1	Year 2	Year 3	Year 4	Year 5	Year 6 to End of Flood
Production						
Oil, bbl						
Gas, Mcf						
Revenues, dollars						
Investments, dollars						
Operating costs, dollars						
Overhead costs, dollars						
Taxes,* dollars						
Cash flow,** dollars						

*Includes ad valorem (property), severance (production), and windfall profit taxes.
**Equals revenues minus sum of investments, operating and overhead costs, and taxes.

TABLE 9.4—ECONOMIC SUMMARY EXAMPLE, INCREMENTAL ECONOMICS (BFIT)			
	Oil Price ($/bbl)		
	10	15	20
Gross oil,* bbl	2,531,000	2,978,000	2,993,000
Gross investment, 10^3 dollars	1,141	1,141	1,141
Payout, years	0.79	0.61	0.51
Profit/investment ratio	9.0	18.2	28.0
Investor's rate of return, %	>100	>100	>100
Present-worth value, 10^6 dollars			
at 8% discount factor	7.1	13.5	20.1
at 10% discount factor	6.5	12.3	18.1
at 12% discount factor	6.0	11.1	16.5
at 15% discount factor	5.3	9.8	14.4

*This changes with oil price because of the change of the point of economic abandonment.

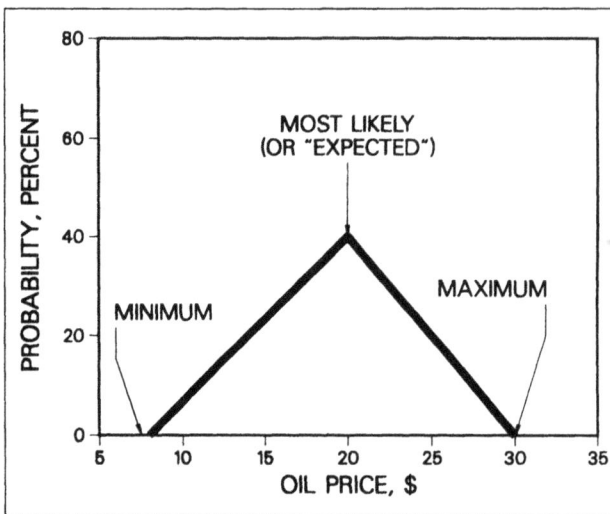

Fig. 9.1—Triangular probability distribution.

The three major cost categories for the waterflood are the costs of (1) obtaining and treating the water, (2) the pump station and injection system expenses, and (3) production operations. Estimating these costs is best done with comparable values from waterfloods, preferably in the same or similar zones in nearby fields. Data obtained in this manner should be discussed with company operation and production engineering personnel to determine what modifications are indicated. These personnel may also be the best source for an operating cost estimate if no local waterfloods exist. Unit costs are frequently the most useful form for estimates. Generally, parameters such as the cost per barrel of source water produced, water treated, water injected, and fluid produced can be applied to the predicted waterflood volumes to estimate operating costs for the waterflood. As noted above, these expense estimates must be increased to account for inflation in future years.[1]

The engineer must ensure that the added costs of processing more water/oil emulsions and sour-gas (H_2S) operations are included in the estimate of operating costs, as appropriate. If comparable waterflood data are being used, all factors tend to be included. If a "from-scratch" estimate is being developed, such factors as increasing water cuts, sour gas, increased corrosion and scale, increased regulatory activity, well stimulations, and increased workover activity become very important. As a general rule, the engineer should guard against underestimating, not overestimating, future operating costs.

When the waterflood operating expenses (including nonincome taxes and overhead) have been estimated, the incremental costs are calculated by subtracting the cost of continued primary production on a year-by-year basis. Having estimated the investment and incremental expense of the waterflood and knowing the waterflood production and revenue predictions, the engineer is now prepared to make an economic evaluation.

9.3 Calculations and Presentation

Generally, the best method of presenting an economic evaluation is in the form of tabular data showing the cash flow of the waterflood project by year and listing various criteria that can be developed from these data. Appendix C shows some of the more common parameters, their definitions, and how to calculate their values. Management will normally use all, or at least several, of these criteria in deciding whether to proceed with the project. Seba's[2] paper is an excellent reference on determining project profitability. Stermole[3] and Megill[4] are also recommended general references for economic evaluations.

A tabular presentation will usually include columns of data with headings as shown in Table 9.3. Discounted cash flow[5] calculations (Appendix C) are frequently included in this type of table. The values shown can be the "gross" numbers, values for the entire project, or "net" numbers, gross values multiplied by a company's working interest. A summary of the economic parameters is also advisable. This table or listing should include (see Appen-

dix C) present-worth values, rate of return on investment, total incremental oil, total investment, payout, and any other criteria required by the waterflood engineer's management. An example of such a listing is shown in Table 9.4 for the same Wyoming waterflood shown in Table 9.2. Table 9.5 shows an example of the calculations that are summarized by tabulations such as those in Tables 9.3 and 9.4. Many companies have calculator or computer programs[6-9] that will perform the calculations done by hand in Table 9.5. Such programs frequently include price and inflation forecasts, as well as automatic tax calculations. The engineer should seek out such time-saving programs before attempting manual calculations. As shown in Appendix C and Table 5, however, these calculations are not difficult, although time-consuming.

Each company will set its own minimum value for such criteria as payout, rate of return, profit-to-investment ratio, and investment efficiency. Management may also use other parameters, such as short-term cash flow or book profit. The engineer should include all required and desirable factors in the presentation form.

9.4 Sensitivities

Table 9.4 shows the sensitivity of the economic parameters to oil price. In many cases, it is recommended that the engineer calculate the effects of changes in variables with high value uncertainty.[10] Factors that usually have a significant probability of variation include rate predictions, oil prices, inflation, operating costs, and investments. The economic parameters are calculated for different values of these variable factors while holding all other conditions constant. As shown in Table 9.4, for example, changing the oil price from $10 to $20/bbl [$63 to $126/m^3$] almost triples the present-worth values. However, payout is much less sensitive to oil price because doubling the oil price ($10 to $20/bbl [$63 to $126/m^3$]) only decreases payout from 9 to 6 months.

In calculating sensitivities, the range of variation for the different factors must be estimated to ensure that the most probable outcome is properly bracketed. Discussions with the reservoir engineers may reveal that the accuracy of the oil-rate prediction is considered to be within 20% and the timing of the peak (or initial) response time is within 1 year. These values could then be expected to be the limits (or brackets) of the probable oil rate from the waterflood. Sensitivity to oil-rate prediction variations can then be calculated with these limiting conditions. Such factors as price, costs, inflation, or investments can be varied similarly and the resultant economic factors tabulated for the management summary.

A range of values gives a better representation of the "real world" outcomes of waterfloods. Greater uncertainty in the variables will lead to the calculation of wider ranges in the economic results. For most cases, a simple probability distribution, such as the triangular one shown in Fig. 9.1, will be a reasonably good representation of the available data. In this case, the engineer has estimated that the oil price will not be less than $8/bbl [$50/m^3$] or more than

TABLE 9.5—EXAMPLE ECONOMIC CALCULATION

Revenues

Year	Incremental Production* Oil (10³ bbl)	Gas (MMcf)	Price Forecast Oil ($/bbl)	Gas ($/Mcf)	Revenues (10³ dollars) Oil	Gas	Total
1	100	250	15	1.50	1,500	375	1,875
2	250	100	19	1.50	4,750	150	4,900
3	300	10	20	1.50	6,000	15	6,015
4	275	0	20	—	5,500	0	5,500
5	200	0	20	—	4,000	0	4,000
6	160	0	20	—	3,200	0	3,200
7	120	0	25	—	3,000	0	3,000
8	90	0	28	—	2,520	0	2,520
9	60	0	30	—	1,800	0	1,800
10	30	0	32	—	960	0	960
11	20	0	36	—	720	0	720
12	15	0	39	—	585	0	585

Expenses

Initial investment for waterflood is $6,000,000. All values include inflation.

Year	Operating Costs (10³ dollars)	Overhead (10³ dollars)	Taxes** (10³ dollars)	Total Costs (10³ dollars)
0				
1	400	100	188	688
2	1,800	360	490	2,650
3	2,100	420	601	3,121
4	1,925	385	550	2,860
5	1,500	300	400	2,200
6	1,280	256	320	1,856
7	1,200	240	300	1,740
8	1,080	216	252	1,548
9	900	180	180	1,260
10	600	120	96	816
11	500	100	72	672
12	400	80	70	550

Cash Flow

Year	Revenue (10³ dollars)	Costs (10³ dollars)	Investment (10³ dollars)	Undiscounted Cash Flow (10³ dollars)
0	—	—	−6,000	−6,000
1	1,875	688	—	1,187
2	4,900	2,650	—	2,250
3	6,015	3,121	—	2,894
4	5,500	2,860	—	2,640
5	4,000	2,200	—	1,800
6	3,200	1,856	—	1,344
7	3,000	1,740	—	1,260
8	2,520	1,548	—	972
9	1,800	1,260	—	540
10	960	816	—	144
11	720	672	—	48
12	585	550	—	35
			Total	9,114

Discounted Cash Flows

Year	Undiscounted Cash Flow	Discount Factors At 8%	At 10%	At 12%	At 15%	Discounted Cash Flow At 8%	At 10%	At 12%	At 15%
0	−6,000	1.00	1.00	1.00	1.00	−6,000	−6,000	−6,000	−6,000
1	1,187	0.96	0.95	0.94	0.93	1,140	1,128	1,116	1,104
2	2,250	0.89	0.86	0.83	0.80	2,003	1,935	1,868	1,800
3	2,894	0.82	0.78	0.74	0.69	2,373	2,257	2,142	1,997
4	2,640	0.76	0.71	0.66	0.59	2,006	1,874	1,742	1,558
5	1,800	0.70	0.64	0.58	0.51	1,260	1,152	1,044	918
6	1,344	0.64	0.58	0.52	0.44	860	780	699	591
7	1,260	0.59	0.52	0.46	0.38	743	655	580	479
8	972	0.55	0.47	0.41	0.33	535	457	399	321
9	540	0.51	0.43	0.36	0.28	275	232	194	151
10	144	0.47	0.39	0.32	0.24	68	56	46	35
11	48	0.43	0.35	0.28	0.21	21	17	13	10
12	35	0.40	0.32	0.25	0.18	14	11	9	6
				Present-worth value:		5,297	4,554	3,851	2,969

Investment efficiency at 12% = 3,851/6,000 = 0.64
Investor's rate of return, % 28.9
Payout at 12%, years 3.50
Profit-to-investment ratio = 15,114/6,000 = 2.52

*After royalty payment.
**Production or property, not income taxes.

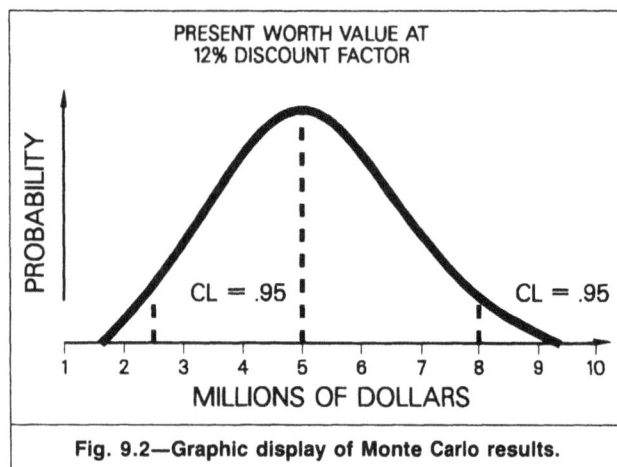

Fig. 9.2—Graphic display of Monte Carlo results.

might have a 95% confidence range from $4,200,000 to $6,000,000. These graphical displays give management an instant "eyeball" estimate of the risk and probable outcome of investing the company's money in the waterflood. The engineer should consider such graphs as well as tabulations of the results as part of the presentation for waterflood project economics calculated with the Monte Carlo technique.

This discussion is not a comprehensive description of economic analysis theory and practice. Numerous references (including those cited) provide an in-depth study of the subject. This chapter can only cover the basics, but many waterflood projects can be adequately analyzed with these techniques. In-house company policies may provide or dictate other procedures. The engineer should use all appropriate information on this subject to provide an accurate economic evaluation.

$30/bbl [$189/m^3] and that the most likely value (with a 40% probability of occurrence) will be $20/bbl [$126/m^3]. Such distributions will yield the limits for each variable that can be used to calculate the sensitivity of the economic parameters to that variable. It is normally unnecessary to formalize the process to the degree of actually drawing the distributions; a mental estimate will suffice. The process of varying one factor while holding all others constant is also sufficient for calculating sensitivities for most waterflood projects.

In more complex cases, the engineer may need to use the Monte Carlo technique to allow a more accurate evaluation of the uncertainty in the project. Several excellent papers[10-13] discussing the use of this procedure are recommended for further information on this subject. The basic procedure is fairly simple, but very time-consuming if done manually. The Monte Carlo technique can be performed quite rapidly with a computer program. The process is as follows: (1) set up probability distributions for every significant variable in the evaluation, (2) pick a value at random from within each distribution (usually a random-number generator is used here), (3) calculate the cash flow for the picked values, (4) calculate the desired economic parameters, and (5) repeat steps 1 through 4 many times (approximately 50 is usually more than sufficient). The resultant distribution of parameter values can be analyzed statistically to yield a mean value for the parameter and a range about that mean. The Monte Carlo method has the advantage that it takes into account the simultaneous variation of the different factors. However, the effort required is normally necessary in only the larger, more complex waterflood projects. The results of a Monte Carlo technique can be presented as a tabulation of the means and ranges for the different parameters or as distributions. Fig. 9.2 shows an example of the distribution display. In this case, the Monte Carlo results showed a fairly broad distribution with a mean of $5,000,000 and a range from $2,500,000 to $8,000,000 at 95% confidence limits; i.e., one can be 95% certain that the actual value will fall within that range. A narrow (or tight) distribution around this mean

References

1. Davidson, L.B.: "Investment Evaluation Under Conditions of Inflation," *JPT* (Oct. 1975) 1183–89.
2. Seba, R.D.: "Determining Project Profitability," *JPT* (March 1987) 263–71; *Trans.*, AIME, **287**.
3. Stermole, F.J.: *Economic Evaluation and Investment Decision Methods*, fifth edition, Investment Evaluations Corp., Golden, CO (1984).
4. Megill, R.E.: *An Introduction to Exploration Economics*, Petroleum Publishing Co., Tulsa, OK (1971).
5. Reynolds, F.S.: "Discounted Cash Flow as a Measure of Market Value," *JPT* (Nov. 1959) 15–19.
6. Santos, R.: "Calculator Program Finds Present Value and Rate of Return on Investment Opportunities," *Oil & Gas J.* (Dec. 21, 1981) 62–68.
7. Jones, R. and Maxwell, R.L.: "Hand-held Calculator Program Gives Economic Evaluation," *Oil & Gas J.* (Feb. 14, 1983) 102–08.
8. Teague, N.: "Calculator Program Finds Cumulative DCF in One Computation," *Oil & Gas J.* (March 7, 1983) 105–12.
9. Coffey, L.F.: "Spreadsheet Program for Handheld Has Variety of Uses, Parts 1, 2 and 3," *Oil & Gas J.* (March 26, 1984) 133–34, (April 2, 1984) 141–42, and (April 9, 1984) 99–102.
10. Walstrom, J.E., Mueller, T.D., and McFarlane, R.C.: "Evaluating Uncertainty in Engineering Calculations," *JPT* (Dec. 1967) 1595–1603; *Trans.*, AIME, **240**.
11. Evers, J.F.: "How To Use Monte Carlo Simulation in Profitability Analysis," paper SPE 4401 presented at the 1973 SPE Rocky Mountain Regional Meeting, Casper, WY, May 15–16.
12. Behrenbruch, P., Turner, G.J., and Buckhouse, A.R.: "Probabilistic Hydrocarbon Reserves Estimation: A Novel Monte Carlo Approach," paper SPE 13982 presented at the 1985 SPE Offshore Europe Conference, Aberdeen, Sept. 10–13.
13. Garibaldi, C.A. and Iskander, F.F.: "A Simulation Approach To Viability Analysis," paper SPE 14157 presented at the 1985 SPE Annual Technical Conference and Exhibition, Las Vegas, Sept. 22–25.

SI Metric Conversion Factors

bbl × 1.589 873	E−01	= m^3
ft^3 × 2.831 685	E−02	= m^3

*Conversion factor is exact.

Chapter 10
Government Regulation

10.1 Introduction

This chapter covers compliance with regulations for waterflood operations. The major areas covered are the regulations promulgated by the U.S. federal government that specifically control the source and underground injection of water in wells and the state permits needed for waterflooding in the U.S. The main emphasis is on environmental rules to protect air and water resources. In the U.S., most state and local regulatory agencies patterned their laws after federal examples; hence, the general subject matter below will likely also apply to state and local agencies. Regulations in other countries tend to be similar to U.S. laws. This chapter includes only the rules for U.S. operations. Waterflood engineers designing a project in another country should consult the appropriate government agency or agencies to determine the required permits.

This chapter does not include procedures to gain permits to drill wells, to lay lines, or to perform other routine oilfield operations. Instead, the requirements for permits to produce and inject water under various U.S. federal and state regulations will be covered in some detail. A list of various federal laws that may apply to a waterflood installation is shown in Table 10.1. A list of the pertinent state regulatory agencies for the major oil-producing states is shown in Appendix D. The U.S. DOE handbook[1] on the environmental regulations for enhanced recovery is recommended as a resource for permitting waterfloods. Table 10.2 summarizes the contents of this handbook. All published materials (including this monograph) concerning environmental or other government regulations may be out of date. The waterflood engineer should consult the latest available information on this subject. Discussions with the government agencies involved before beginning the permit process frequently can greatly shorten the time required for approval. The engineer should consider that some requirements or stipulations imposed by the permitting agencies may not be justifiable or reasonable for the specific conditions of the project. Using personnel (in-house or a consultant) experienced in working with these government departments to make the initial contacts may facilitate negotiating a solution that satisfies the regulatory needs at a more acceptable cost and effort.

10.2 Air Pollution Control

A waterflood project may need an air quality permit. Normally, this will be required if the waterflood design involves a sufficient number of internal-combustion or turbine engines that the exhaust emissions exceed the allowable limit. This limit varies widely, depending on the regulatory agency and the ambient air quality in the area. The U.S. Environmental Protection Agency (EPA) requires Prevention of Significant Deterioration (PSD) permits for all "major sources." Major sources are defined as those facilities that emit more than 250 tons/yr [227 Mg/a] of air pollutants (except fossil-fuel boilers and very large oil storage tanks, which are limited to 100 tons/yr [91 Mg/a]). If a source is classified as major, a permit is required for any increase in emissions. For other sources (including new sources), a PSD permit is needed whenever the total emissions are expected to exceed this limit. In many areas, either state or local agencies have been granted primacy by the EPA. This means that the EPA has delegated its regulatory powers to those

agencies because the rules they administer meet or exceed the federal standards. In areas such as California, the limits are set much lower (25 tons/yr [23 Mg/a]).

Obtaining a PSD permit is not easy and is very time-consuming. Unless already available, ambient air quality and meteorological data must be collected on site for 12 months to establish a baseline. The monitoring program must be designed to the standards set by the EPA. The effect of the applicant's emissions on ambient air quality must be predicted, normally by dispersion modeling techniques. Obtaining the necessary data, performing the modeling, and filing for the PSD permit are very complex processes. Various consulting firms or in-house environmental groups are available to carry out the needed work. The waterflood engineer should discuss the requirements with in-house or outside experts to determine the need for a PSD permit. If one is necessary, the engineer should use the available resources or outside companies to obtain the permit.

10.3 Water-Injection Permits

The majority of waterflood installations are not large enough to exceed the emissions limitations and hence do not require permits for air quality control. However, all water-injection wells must now be permitted, as specified by the Safe Drinking Water Act. The federal regulations[2] designate different classifications of wells. Underground injection wells related to oil or gas production and injecting produced or source water for secondary recovery or disposal are in Class II. In most major oil-producing states (except Pennsylvania and Montana), the state agencies have taken primary responsibility for administering the Underground Injection Control (UIC) program of the EPA. These state regulatory programs are directly modeled on the EPA requirements, so only the EPA program is covered here.

10.3.1 Secondary Recovery Permits. In virtually all states, a secondary recovery project, such as a waterflood, will require an overall permit. The basic purposes are to protect the rights of offset operators and royalty owners and to prevent waste of the hydrocarbon resource. The permitting agency will normally require information from the reservoir study showing the recovery of additional oil by flooding (or, at least, no loss of hydrocarbons). It will also be necessary to demonstrate that no offset operator or owner will be damaged by the waterflood. Notification of all interested parties in the area, including the public, is usually required. A hearing, at which evidence is presented by all who wish to do so, is frequently held by the agency before a permit is issued.

In Texas, for example, the Texas Railroad Commission (TRRC) has jurisdiction over waterfloods. State Rule 46 governs the injection of fluid into productive reservoirs. Permits may be granted when "injection will not endanger oil, gas, or geothermal resources or cause the pollution of fresh water strata."[3] Copies of the application for a permit to inject must be sent to the surface owners, offset operators, and county and city clerks and must be published in a newspaper in general circulation in the area. These notices must be made before or at the time the permit application is filed with the TRRC. If any protests are made or if the director of UIC de-

<table>
<tr><td>

TABLE 10.1—MAJOR U.S. LEGISLATIVE ACTS PERTINENT TO WATERFLOOD OPERATIONS

Natl. Environmental Policy Act
Clean Air Act
Resource Conservation and Recovery Act, including the
  Hazardous and Solid Waste Amendments of 1984
Clean Water Act
Safe Drinking Water Act
Coastal Zone Management Act
Rivers and Harbors Act
Toxic Substances Control Act
Noise Control Act
Endangered Species Act
Water Resources Planning Act
Marine Protection and Sanctuaries Act
OCS Lands Act
Federal Oil and Gas Royalty Management Act
Occupational Safety and Health Act
Comprehensive Environmental Response, Compensation
  and Liability Acts, including the Superfund
  Amendments and Reauthorization Act of 1986

</td></tr>
</table>

<table>
<tr><td>

TABLE 10.2—CONTENTS OF U.S. DOE HANDBOOK

General Guidelines for Complying With Environmental
  Laws and Regulations
    Obtaining a permit
    Compliance and enforcement
    Federal/state relations

Air Pollution Control
    Clean Air Act
    State implementation
    New source permits

Water Pollution Control
    Clean Water Act
    Wastewater discharge
    Natl. Pollution Discharge Elimination System permit
    Spills and spill prevention
    Hazardous substances
    Corps of Engineers permits—dredge and fill and
      construction

Underground Injection Control

Hazardous Waste Management

Major Federal Laws Affecting Siting or Operation of
  EOR Facilities

State Regulations: Alabama, California, Colorado,
  Illinois, Kansas, Kentucky, Louisiana, Mississippi,
  Montana, New Mexico, Ohio, Pennsylvania,
  Texas, West Virginia, and Wyoming

</td></tr>
</table>

cides it is appropriate, a public hearing will be called. The information to be included with the application includes reservoir and fluid data, such as name, lithology, depths, pressures, FVF, gas-cap information, productive area, structural geology, porosity, permeability (horizontal and vertical), pay thickness, oil gravity, fluid saturations, and oil viscosity. A description of the production history must include the number of wells, current producing rates, cumulative production, stage of depletion, and total production by year or month since the reservoir was discovered. The injection project must also be described, stating such data as additional oil recovered, expected residual oil saturation, injection pattern and spacing, number of injection wells, source of injection water, total injection rate, maximum injection rates for each well, and maximum injection pressure. Values for these parameters may be developed as discussed in Chaps. 3 through 5. A full description of the completion and mechanical condition of each injection well must be appended to the application. If fresh water is to be injected as part of the project, then it must be proved that no substitute is available that is economically and technically feasible to use. Most oil-producing states require essentially identical information.

In many companies, this type of permitting effort is performed by reservoir engineers with the help of the waterflood engineer as needed. If the waterflood engineer must head this process, then close coordination with reservoir personnel is essential. It is also suggested that early contact and discussions with the appropriate regulatory agency be held. This can clarify the application requirements and speed the approval of the waterflood. All states (see Appendix D) have published their regulations on this subject. The engineer should obtain copies of the rules for the pertinent states.

10.3.2 Federal Injection Permits. The EPA application for a Class II injection well permit is EPA Form 4 (7520-6). The information needed is the facility name and address; owner/operator's name and address; land ownership status (federal, state, private, etc.); well status, including numbers of existing/proposed wells, class and type of wells, and name(s) of field(s); location of well(s); and a certification that the applicant is authorized to submit the application on the behalf of the operator(s). Additionally, the EPA requires the following attachments.

1. Area of review calculation, if different from statutory limit. The area of review is that area around an injection well within which the applicant must supply various items of information on each and every well. The states normally set a statutory limit of a ¼- or ½-mile [0.4- or 0.8-km] radius for the area of review. The engineer should refer to the statutes of the specific state involved to determine the data required and the limiting radius. Any difference from the statutory limit must be justified in the application; e.g., the limit should be ⅛ mile [0.2 km] in a certain direction because of a permeability pinchout of the reservoir at that distance.

2. Topographic maps of the wells and area showing facilities.

3. Well data and corrective action plans, if needed, for all wells in the area of review.

4. Geologic name(s) and depth(s) of underground sources of drinking water.

5. Geological data on injection and confining zones.

6. Operating data—e.g., injection pressures, rates, water sources, annular fluids, and water analyses.

7. Formation testing program to determine fluid and fracture pressures and physical characteristics.

8. Stimulation program, if any.

9. Injection procedures, including pumps, tanks, manifolds, etc.

10. Construction procedures and details, including casing and cementing program, logging, drilling, and completion (or recompletion) plans. Schematics or other drawings of the well(s) surface and subsurface details should be included.

11. Plans for well failures.

12. Monitoring programs for fluid sampling, pressure changes, rate changes, and other parameters. Frequency of testing should be included.

13. Plugging and abandonment plans (EPA Form 7520-14).

14. Aquifer exemptions.

15. Existing EPA or state permits.

16. Proof of financial resources.

17. Description of business.

The last two can usually be satisfied by a copy of the company's latest annual report to the stockholders, if available.

The purpose of this application and the attached information is to allow the EPA to determine that the injection well does not pose any danger of polluting drinking-water formations. When they have so decided, the EPA will issue a UIC permit, normally for the life of the injection well, with stipulations for additional requirements as pertinent for any particular well. Some state agencies have further requirements beyond those outlined above. For example, Wyoming requires a permit from the State Engineer's office to appropriate water (even produced water) for underground injection. The waterflood engineer should ensure that such additional information is provided for the specific state in which the waterflood is located. The DOE handbook[1] gives further details of the data required.

10.4 Special Problems

If potable water is to be the source water for a flood, then further information will be needed. The engineer should carefully research the available sources and predicted use of fresh water for public consumption, agriculture, industrial, or commercial facilities. As mentioned in Chap. 4, water rights must be obtained. Fresh water is not always a renewable resource, and governments tend to be very protective of the available water. Hence, the waterflood engineer must include the possible difficulty or even impossibility of getting a permit when determining the best source water.

If the waterflood is to be located offshore in federal waters, then the lead agency in approving a project is the Minerals Management Service (MMS) of the Dept. of the Interior. The Coastal Zone Management Act (CZMA) dictates that contiguous states will also become involved. Offshore within state lands, the state agencies will generally have jurisdiction.

In U.S. Outer Continental Shelf (OCS) waters, the MMS requires operators to file plans of production. These are simultaneously submitted to the state that has authority under the CZMA. A waterflood would constitute a major change in the plan and would therefore require a new plan to be proposed and approved. The state will influence the approval by the MMS in direct proportion to the effects the project will have on and near the shoreline. For example, if the injection or production facilities must be partially or wholly on land, the state will likely have veto power over the entire project. The waterflood engineer should be familiar with the general rules, known as OCS Orders 1 through 12, as well as site-specific orders for certain localities—e.g., offshore Alaska. The engineer should be prepared to satisfy the MMS that the design for the waterflood will increase the amount of oil recovered (and hence royalties) in an environmentally sound manner, safely, and without potential threat to public health.

Onshore federal lands require that a waterflood meet very similar standards, but the supervising agency is the Bureau of Land Management. Other agencies, such as the Forest Service, the Bureau of Reclamation, the Corps of Engineers, and the Bureau of Indian Affairs, may also become involved, depending on the ownership status of the land and mineral rights. The permitting process clearly becomes more complex and lengthy as more participants become active. Hence, waterflood projects on Indian land, in wilderness areas, near reservoirs or dams, and in wetlands will require extra efforts to meet the special conditions in such regions.

In some cases, local and state governments have established additional rules with which one must comply. An example is Wyoming's Industrial Siting Act that specifies state approval when the cumulative investment in a facility reaches a designated value. The waterflood engineer should discuss any such laws with local, company engineering, or production supervision personnel, if they are available.

As a general recommendation, the engineer should consult the operators of similar waterfloods in the same government jurisdictions, whether local, state, or federal. The production and engineering personnel of these operators are usually an excellent source of information concerning the agencies and permits needed to start a waterflood project. They may also be able to introduce the engineer to agency personnel who have proved to be helpful and knowledgeable.

10.5 Summary of Procedure

The following general steps are suggested for obtaining the required permits.

1. Define the operation fully and precisely (details may affect laws and permits).

2. Identify the facility site with acceptable alternates.

3. Survey environmental and geographical conditions at the site(s).

4. Identify permits expected to be required, including those from county, city, township, area-wide, tribal, state, and national agencies.

5. Establish direct contact with the permitting agencies.

6. Identify special studies needed to obtain permits. If required, these should be implemented in a timely manner.

7. Submit the permit applications early enough to ensure that the permit will be issued when needed but will not expire before being used.

8. Check the status of the application with informal contacts between designated persons.

Permit processing can be tedious and even frustrating for the waterflood engineer. If adequate time for the procedure has been allowed, however, the emotional distress is greatly reduced. The engineer should strive at all times to maintain a cordial and professional relationship with the personnel at the permitting agencies. These people can be of great assistance in expediting permits and are more likely to be helpful if the requestor has consistently been on good terms with them. The regulators' job has been specified by law as beneficial to the public interest. This fact should be kept clearly in mind by permit applicants. If all parties are cooperative and act professionally, the permitting process will proceed smoothly and the approvals will be timely.

References

1. Wilson, T.D. and Kendall, R.F.: *Environmental Regulations Handbook For Enhanced Recovery—1983 Update*, DOE/BC/10355-1, U.S. DOE, Bartlesville, OK (Oct. 1983).
2. Wasicek, J.J.: "Federal Underground Control Regulations and Their Impact on the Oil Industry," *JPT* (Aug. 1983) 1409-11.
3. *Underground Injection Control Reference Manual*, Oil and Gas Div., Texas Railroad Commission, Austin (Jan. 1985).

Chapter 11
Cooperative Agreements

11.1 Need for Cooperative Agreements

Divided ownership is common in most oil reservoirs in the U.S. as a result of competition for leases. Early in the industry's history, waterfloods typically were operated on individual leases without considering the effect the injected water could have on offset properties in the same reservoir. This approach pushed oil across lease lines and "watered out" other operators' wells. Currently, a waterflood must follow certain legal processes designed to promote conservation of resources and to protect the individual rights of all interest owners. All the oil- and gas-producing states have adopted conservation laws, and most states have provided for either voluntary or statutory cooperative agreements. The U.S. federal government has similar provisions in the Mineral Leasing Act of Feb. 25, 1920, which allows federal lessees to adopt and operate a cooperative or unit plan of development collectively. These agreements are administered through the Bureau of Land Management and the Minerals Management Service. Commitments can be made between operators to cooperate or to agree to unilateral departures in field operation, but the rights of all parties must be considered. Two methods of protecting the rights of all parties are unitization and lease-line or cooperative agreement(s). This chapter briefly describes both solutions and their effects on a waterflood project.

11.2 Unitization

Unitization is the most common method of protecting the rights of all in implementing and operating a water-injection project for an oil reservoir owned by several parties. Under unitization, each owner allows his or her particular tract to be combined (or pooled) with other tracts in exchange for a fractional interest in the production from the entire reservoir or that part which is unitized. The resulting "unit" is an entity that is not internally limited by the original property configurations. In effect, property lines do not exist within the unit, thus allowing the most favorable pattern arrangement of injection wells, efficient field development, and cost-effective operating practices. Ideally, the unit should include the entire reservoir to be waterflooded. It may also need to encompass the area downdip of the oil-producing zone when an edge pattern is expected.

The operator with the largest ownership in a unit is usually designated the "unit operator." The unit operator conducts all field operations under such restrictions as the agreement imposes. The company (or individual) that will most likely have the largest interest normally initiates efforts toward unitization. However, there is no restriction to prevent another company or individual from taking the lead.

11.2.1 First Steps Toward Unitization. If unitization appears to allow the most economically successful waterflood, the initiating company will invite representatives of the potential working-interest owners to meet to discuss unitization procedures. The representative of the initiating company will act as temporary chairman. The results of a waterflood feasibility study are usually presented and a recommendation made to proceed with formation of a unit. If there is sufficient agreement, the company representatives present may decide to form an operators' committee (or a working-interest owners' committee) for the proposed unit. Other committees may also be formed as needed. Negotiations leading to unit formation can

then begin. Details on how to conduct the negotiations are not covered in this monograph. If the initiating company has prepared a complete study acceptable to the working-interest owners, the subsequent effort is greatly simplified. This type of planning is discussed in Chap. 2. If this has not been done or if major revisions are necessary, an engineering committee may be formed.

11.2.2 Unit Area and Unitization Parameters. Whenever possible, the unit area should include all productive acreage and all usable wells in the reservoir. The unit boundary usually follows fractional survey lines outside the productive limit (Fig. 11.1). With edge injection, the limit will lie outside the proposed injection wells. An approved base map showing all tracts (separate ownership leases or parcels) is made.

The theory of unitization holds that each interest owner in the unit must receive an equitable fraction of the production from the entire reservoir instead of that which is produced only from the individual owners' specific lease(s). Only through this equitable fraction can the rights of all parties, including those whose wells are to be used for injection purposes, be protected adequately. Simply stated, the equitable fraction, as determined by the participation formula, provides that each owner shall receive an interest in the unit that reflects the relative value of his or her tract(s) to the other tracts contributed to the unit. Participation formulas are developed that use numerous parameters to ensure equitable participation for each owner. Some common parameters are (1) productive acreage, (2) acre-feet of effective reservoir, (3) recoverable oil in place, (4) producible reserves, (5) value of future reserves, (6) bottomhole pressure, (7) current production, (8) cumulative production, (9) number, completion design, and condition of wells, (10) acre-feet times porosity, and (11) acre-feet times permeability.

A multiple-phase participation formula is not unusual. The first phase may have a certain equity formula until the calculated primary reserves are recovered. Then a second phase would use a different equity formula for the remaining waterflood reserves. For example, the owners might negotiate a two-phase participation formula that gives a Phase 1 interest based on 50% remaining primary reserves and 50% current production. Phase 2 participation might be based on 30% current production, 10% acre-feet, 30% waterflood reserves, 20% number of usable wells, and 10% production acreage. For three owners in the reservoir, the parameter values calculated by a joint engineering committee are assumed to be as shown in Table 11.1. Owners A, B, and C would have the Phase 1 and Phase 2 working-interest participations calculated in Table 11.2. In this example, Owner B changes working interest only slightly between the two phases, but Owners A and C experience significant variance.

The participation formula negotiations are obviously critical to the unitization effort. Small changes can have a significant effect on a company's economic status. For example, at the time of unitization, the Prudhoe Bay Unit in Alaska had a present-worth value discounted at 12% of several tens of millions of dollars for a 0.1% (0.001 working interest) change in equity. Thus, even for a large company, a small variation in its working interest at Prudhoe caused by changes in the equity formula or parameters was very important.

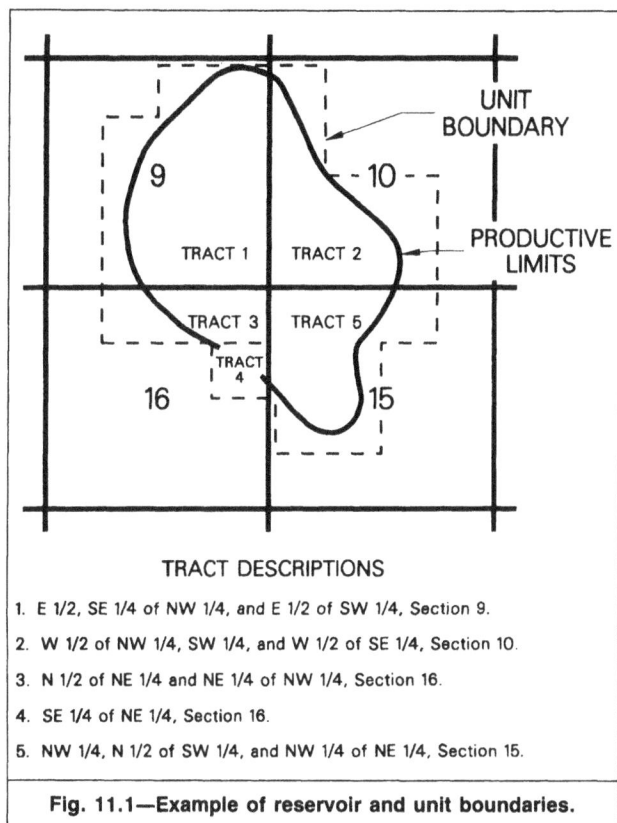

TRACT DESCRIPTIONS

1. E 1/2, SE 1/4 of NW 1/4, and E 1/2 of SW 1/4, Section 9.

2. W 1/2 of NW 1/4, SW 1/4, and W 1/2 of SE 1/4, Section 10.

3. N 1/2 of NE 1/4 and NE 1/4 of NW 1/4, Section 16.

4. SE 1/4 of NE 1/4, Section 16.

5. NW 1/4, N 1/2 of SW 1/4, and NW 1/4 of NE 1/4, Section 15.

Fig. 11.1—Example of reservoir and unit boundaries.

	Owner A	Owner B	Owner C
TABLE 11.1—EXAMPLE: ASSUMED SHARE OF PARAMETERS BY OWNER			
Current production	0.25	0.40	0.35
Remaining primary reserves	0.28	0.36	0.36
Reservoir acre-feet	0.33	0.37	0.30
Waterflood reserves	0.38	0.35	0.27
Number of usable wells	0.20	0.40	0.40
Productive Acreage	0.26	0.39	0.35

agree that the terms of the unit documents, including the participation formula, are fair and equitable. In Arkansas and Louisiana, the prescribed majority is 75%, while in Oklahoma it is 63%. Some states allow the regulators to call a hearing, on their own motion or on the application of any interested person, to consider a unit operation. The agency may order the formation of a unit if the required percentage of interest owners agrees and conservation and legal requirements are also met. Those who do not voluntarily agree to unitize are compelled to join and their interest is effectively committed to the unit subject to available legal remedies.

Regulators have other options as well. The Norge Marchand Unit,[1] formed under the laws of Oklahoma, is an example of a compulsory unit that had problems during and after formation. In Aug. 1971, even before reservoir development was complete, the working-interest owners met to discuss unitization. In March 1973, the Oklahoma Corporation Commission imposed an allowable reduction on the field to conserve reservoir energy while unitization efforts continued. In May 1973, participation was negotiated and agreed to by the owners of just over 63% of the working interest. At the same time, the unit operator was selected. The plan of operation for the unit was presented to the Commission in Aug. 1973 and the unit was approved and became effective on April 1, 1974. Litigation against the unit, however, delayed water injection. In Oct. 1974, the Oklahoma Corporation Commission severely restricted unit oil production to 1,000 B/D [160 m³/d], down from 7,000 B/D [1100 m³/d]. A Supreme Court order in Feb. 1975 upheld the unit and permitted the initiation of water injection. In July 1975, almost 4 years after the first unitization attempts, a planned increase in unit production rate of 500 B/D [80 m³/d] per month was started with the goal of reaching the allowable 7,000 B/D [1100 m³/d] in 12 months.

As the number of owners increases, the unitization process becomes more complicated as each tries to obtain his "fair share" of the equity. Operators should strive for a simple formula that does not require interpretation. Unfortunately, negotiations are frequently so difficult that complex equations are necessary.

A voluntary unit is a term applied to a unit in which only those interests voluntarily committed to the unit participate in the unit. In states that do not have compulsory unitization or pooling statutes, all units are this type. The statutes of the state in which the property is located must be reviewed to determine whether the unit will have to be a voluntary unit. A majority of producing states have some type of compulsory unitization regulations.

11.2.3 Compulsory or Involuntary Unitization. Most oil- and gas-producing states have laws authorizing their oil and gas regulatory agencies to compel unit operations. The statutes for the state in which the project is located should be reviewed, however, to determine whether that state provides for compulsory, or involuntary, unitization and the specific requirements for invoking that power. Generally speaking, for an application for compulsory unitization to be considered by the state regulatory body, a certain, defined majority of the working-interest owners and royalty owners involved must voluntarily agree that a unit is necessary to prevent waste, to protect individual rights, and to increase economically the ultimate hydrocarbon production. In addition, this defined majority must

11.2.4 Unit Agreement and Unit Operating Agreements. Model forms of a unit agreement and a unit operating agreement have been published by the API and are in widespread use in the industry. There are two sets of these forms, one for voluntary unitization and another for statutory (compulsory or involuntary) unitization. The model forms frequently require alteration to suit specific units. Some attorneys prefer slightly altered wording or add special provisions that apply only to the unit in question. Units have been formed with agreements prepared by simply filling in the blanks on the model forms. Other model forms have been developed and are in use. Additionally, some states, notably Oklahoma, currently require that the unit agreement and the unit operating agreement

	TABLE 11.2—EXAMPLE CALCULATION OF UNIT PARTICIPATION BY OWNER	
Owner	Phase 1	Phase 2
A	(0.5 × 0.25) + (0.5 × 0.28) = 0.265	(0.3 × 0.25) + (0.1 × 0.33) + (0.3 × 0.38) + (0.2 × 0.2) + (0.1 × 0.26) = 0.288
B	(0.5 × 0.4) + (0.5 × 0.36) = 0.380	(0.3 × 0.4) + (0.1 × 0.37) + (0.3 × 0.35) + (0.2 × 0.4) + (0.1 × 0.39) = 0.381
C	(0.5 × 0.35) + (0.5 × 0.36) = 0.355	(0.3 × 0.35) + (0.1 × 0.3) + (0.3 × 0.27) + (0.2 × 0.4) + (0.1 × 0.35) = 0.331

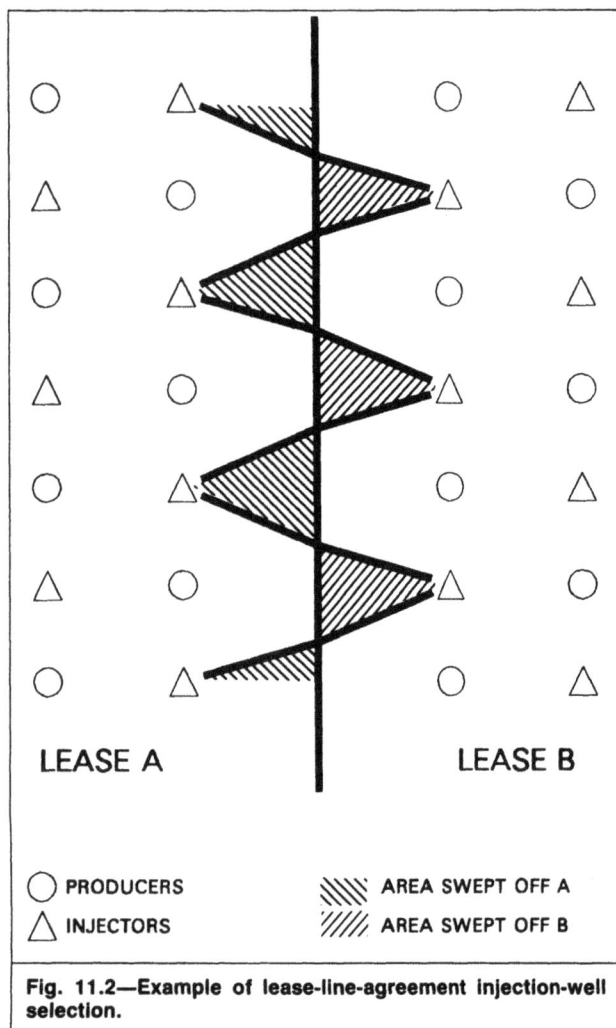

PRODUCERS   AREA SWEPT OFF A
INJECTORS   AREA SWEPT OFF B

Fig. 11.2—Example of lease-line-agreement injection-well selection.

Exhibit C, unit participation; Exhibit D, accounting procedure; and Exhibit E, insurance provision.

Other sections sometimes included are uncommitted royalty, construction of plants and special facilities, multiple completions, and action before effective date. Exhibits A and B are identical to those in the unit agreement. Exhibit C is a schedule showing unit participation of each tract and the total unit participation of each working-interest owner. Exhibit D is the accounting procedure applicable to unit operations. This exhibit states the manner in which the unit operator is compensated for his work; how the owners will be billed; what may be charged to the working interest (both direct and indirect charges); allowable overhead rates; basis of charges to the joint account; disposal of materials; basis of pricing material transferred from joint account; inventories; and audits. The form normally used for this exhibit is known as "COPAS," an acronym for the Council of Petroleum Accountants Socs. of North America, who developed the standard. This form is periodically updated. Older units may use a "PASO" form, originally recommended by the Petroleum Accountants Soc. of Oklahoma and Texas. Exhibit E is a statement of the types of insurance and the amounts that the operator will carry for the benefit of the joint operations. The unit operating agreement is signed by the working-interest owners, but not the royalty holders (unless they are also working-interest owners). The unit operator has the exclusive right and obligation to conduct unit operations subject to the provisions of the agreement and to instructions from the working-interest owners committee. The operator is charged with the responsibility to conduct operations in a good and workmanlike manner as would a prudent operator in the same or similar circumstances. The operator must keep the working-interest owners informed and may consult them freely. The operating personnel for the unit are employees of the unit operator. Accurate books of account and records of unit operations must be kept and periodic reports supplied. All necessary reports to governmental authorities should be filed. The operator is authorized to make expenditures for the unit up to the stipulated limit in the operating agreement without approval of the working-interest owners. As an employee of the unit operator, the waterflood engineer must work within the restrictions of the unit operating agreement.

11.3 Lease-Line Agreements

Lease-line agreements (also referred to as cooperative agreements) are legal contracts entered into by operators where adjacent leases (usually only two) with different working-interest owners are involved. In some cases, one operator will refuse to become party to a voluntary unitization agreement but will agree to a cooperative operation with the unit using a lease-line agreement. Such an agreement will set forth an equitable plan for a water-injection project, select injection wells, and determine an injection plan along or near the property lines.

Fig. 11.2 shows a common method of alternate-property injection-well selection for line-agreement purposes. Although oil will be moved across the property lines by water injection into wells located near the lines, the gains and losses to each property should be equal. The engineer must attempt to ensure that an equitable exchange of oil reserves occurs under this type of plan so as to preserve individual rights. Injection patterns of this type are not always possible. In some cases, owners compensate each other by cash or other payments to settle differences. The rights of the royalty interests must also be protected in any cooperative agreement, and the royalty owners usually are made signatories.

Lease-line agreements sometimes provide for newly drilled water-injection wells along common property lines. Fig. 11.3 illustrates an example where property owners agree to drill line injection wells and equitably proceed with separate water-injection projects. It is possible that Property Owners A and B in Fig. 11.3 desire to waterflood while Owners C and D do not. If the facts at a hearing show profitable waterflood prediction, conservation of resources, and protection of owners' rights, then some states provide that their regulatory body may give approval to drill the line injection wells. The waterflood can then proceed even without a line agreement with Property Owners C and D.

be combined in a single document referred to as the "plan of unitization." The waterflood engineer must be familiar with the contents and meaning of the unit documents. Below is a list of provisions frequently found in a two-document unit.

The *unit agreement* is the legal document that defines the unit and the rights of all parties. A typical unit agreement includes (1) definitions, (2) exhibits, (3) creation and effect of unit, (4) plan of operations, (5) tract participation, (6) allocation of unitized substances, (7) use or loss of unitized substances, (8) tracts to be included in unit, (9) titles, (10) easements or use of surface, (11) enlargement of unit area, (12) transfer of interest, (13) relationship of parties, (14) laws and regulations, (15) force majeure, (16) effective date, (17) terms, and (18) counterpart.

The unit agreement exhibits include Exhibit A, a list of tracts and tract participation, and Exhibit B, a map of the unit area. The exhibits are important parts of the agreement. These must be accurately prepared and carefully presented (see Fig. 11.1). The document is executed by both working- and royalty-interest owners and is filed for record with the appropriate regulatory agency.

The *unit operating agreement* defines the manner in which the unit shall be operated and establishes the contractual relationship between the working-interest owners. An operating agreement typically includes (1) confirmation of unit agreement, (2) exhibits, (3) supervision of operations by working-interest owners, (4) manner of exercising supervision, (5) individual rights of working-interest owners, (6) unit operator, (7) authorities and duties of unit operator, (8) taxes, (9) insurance, (10) adjustment of investments, (11) unit expense, (12) nonunitized substances, (13) titles, (14) liability, claims, and suits, (15) internal revenue provision, (16) notices, (17) withdrawal of working-interest owner, (18) abandonment of wells, (19) effective date and term, (20) abandonment of operations, (21) execution, and (22) successors and assigns.

The exhibits in the operating agreement include Exhibit A, a list of tracts and tract participation; Exhibit B, a map of the unit area;

Although each operator operates his own wells, the lease-line agreement usually specifies injection rates and pressures for all injectors along the common lines. These rates (or pressures) may prove unacceptable as the waterflood progresses. Changes may be difficult, or even impossible, to implement by mutual approval. If control is ineffective, one operator may try to take advantage of the other. Line agreements should be avoided unless no other solution is available. This type of arrangement is almost invariably less efficient than a unitized operation.

The waterflood engineer usually is not directly involved in the negotiations for the joint line agreement. However, the design of the flood must comply with the contractual commitments imposed by the agreement. Joint-agreement projects are more difficult to operate than reservoirs under unit agreements. The main disadvantages in a cooperative project are the additional cost, lack of flexibility, and inefficiency. Most cooperative contracts permit the unilateral withdrawal of a working-interest owner, which can destroy the effectiveness of the program and bring an end to the cooperation. The success of a cooperative project largely depends on the mutual trust between the parties. Good records must be kept and information exchanged. There must be a willingness to change plans as needed to accomplish an equitable return from the waterflood.

11.3.1 Factors in Lease-Line Injection. The working-interest owners must negotiate the development of a water supply. One solution is for a single operator to build and operate a source-water and a water-injection facility. In this case, that operator may sell water to the other parties on a cost-per-barrel basis. Alternatively, the entire water supply and injection system may be developed as a joint project, with investment and operating costs shared. If each operator has developed individual water systems, then each might be responsible for supplying a specified amount of water to the injection wells.

All parties must agree on the wells to be converted to water injection. New water-injection wells may be needed, and the sharing of the costs and drilling responsibilities must be determined. Every part of the engineering development and operation must be mutually agreed on by all concerned. Because this contract dictates much of the waterflood design, the waterflood engineer should try to be involved in setting the terms before the final signature.

11.4 Regulatory Permits for Unitization and Waterflooding

The unit operator must submit the unit plan to the state regulatory agency for approval. In most states, a notice will be posted to announce a public hearing. At this hearing, which may be before a representative of the regulatory body, the plan of unitization and operation is presented. If opposition exists, the regulatory agency will hear opposition testimony and obtain clarification or answers from the proponents in response. If the agency decides that the plan is in the public interest, protects correlative rights, will increase ultimate recovery, and will prevent waste, an order will be given to approve the unit operation.

For example, the following definite requirements[2] for approval of a unit in Texas are imposed by the Texas Railroad Commission.

1. The agreement is necessary to accomplish the purpose for which it is made.

2. The agreement is in the public interest because it prevents waste and promotes conservation of oil and gas.

3. The rights of all owners, both those signing the agreement and those not assenting to it, are protected under the proposed plan of operation.

4. The estimated cost of the plan does not exceed the value of the estimated increased recoveries.

5. The other methods of secondary recovery are inadequate.

6. The area covered includes only that defined by development.

7. The basis of participation is equitable and the owners of interests in each and every tract have been extended the same opportunity to participate and on the same basis.

After an order approving the unit and waterflood is obtained from the regulatory body, the minimum number of signatures are on the agreements, and the unit agreement is filed at the county offices,

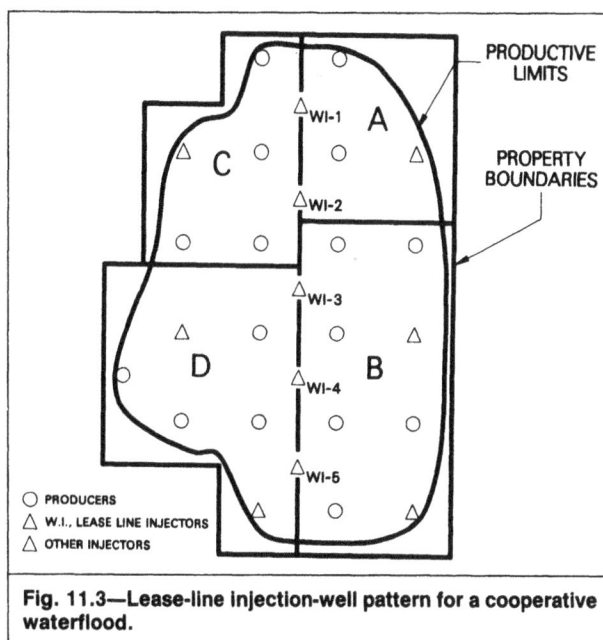

Fig. 11.3—Lease-line injection-well pattern for a cooperative waterflood.

the effective date can be established and operations may start. The effective date is often 7 a.m. of the first day of the month after the agreement is filed for record.

11.4.1 Regulatory Permits for Cooperative Waterflood Operations. The procedure for obtaining approval by the state regulatory body for a cooperative waterflood project is very similar in most respects to getting a permit for a waterflood unit. Similar procedures are necessary, although approval may be more difficult. The agency will frequently request proof that a unit cannot be formed.

11.5 Proration

Although most engineers in most waterfloods will not have to deal with proration (regulatory restriction of production as a conservation measure), a few might. Proration of oil production by state regulatory bodies was instituted at a time when a large oversupply of crude was produced. Oil prices were reduced drastically and wasteful production practices were occurring. The proration method assigns maximum producing rates, also known as allowables, on a well or field basis. Various methods of proration have been used in different states. The engineer should refer to the applicable statutes in the state within which the project is located for the exact technique specified to calculate allowables.

Waterfloods generally are in one of two categories concerning allowables. Waterfloods initiated in depleted reservoirs with production at a low rate are usually allowed to produce at capacity. Capacity means the maximum rate at which they are able to produce, and such projects are therefore free from the effects of proration. The other category is waterfloods that are initiated when there is substantial remaining reservoir energy. These fields may be subject to proration to ensure efficient production practices and the protection of correlative rights. Individual reservoirs that have an assigned maximum efficient rate are not allowed to exceed this rate because waste of underground resources may occur. It is now very rare for a waterflood to be limited by proration for onshore U.S. fields. In large offshore U.S. fields or fields located in other countries, proration or reduced production rates are common, particularly in countries that export large amounts of petroleum. The engineer should evaluate the conditions specific to the waterflood project to determine whether proration is a factor.

References

1. Jowell, J.H. and Haberthur, C.R.: "Norge Marchand Deep Unit," *Pet. Eng.* (Nov. 1976) 21-25.
2. *Texas Oil and Gas Handbook*, R.W. Byram and Co., Austin, TX (Sept. 1985).

Chapter 12
Design Implementation: Building the System

12.1 Design Implementation

This chapter discusses relationships between the waterflood engineer, construction management, contractors, and vendors during the implementation of the waterflood design. In many cases involving small projects or small companies, the engineer may act as the on-site construction supervisor, materials expediter, and trainer. For large projects, the engineer may have only limited interaction with technical personnel and virtually no contact with construction personnel. In these cases, the on-site technical and construction work may be performed entirely by a large engineering/construction company on a contract basis. For medium-to-large projects, many large companies have in-house departments that handle the construction phase. Cost and complexity distinguish projects supervised by such a group from smaller ones that don't require their attention. The experience and background of the waterflood engineer may determine the choice of supervisor in marginal cases. If the in-house group takes over construction responsibilities, then the engineer may be needed as a liaison between the construction team and local operations and engineering management. Gregg[1] discusses the general aspects of project management in a North Sea development.

Because the first case mentioned above (small projects) requires the greatest attention of the waterflood engineer, this chapter will focus mainly on this type of project. Appendix A discusses larger projects not directly supervised by the design engineer. There are many items discussed here and in Chap. 2 that are pertinent to any type of project design implementation. Hill[2] presents a good discussion of the possible pitfalls in project design.

12.2 Plans for Implementation

As outlined in Chap. 2, the finished design includes process flow diagrams, process and instrumentation diagrams, structural and piping layout drawings, electrical one-line drawings, and control schematics. Cost estimates and major equipment listings should also be available at this stage. The engineer must now develop a schedule and a "material take-off." The latter is a complete, detailed list of the materials required by the project, ranging from such major items as pumps, prime movers, and treating vessels to pipe, nipples, and fittings. The schedule may be simple or complex,[3] ranging from dates on a calendar to computer-generated program evaluation and review technique (PERT) diagrams or critical-path diagrams.[4,5] Visual path analysis[6] is another useful technique for planning a schedule. However developed, the schedule, drawings, and material take-off constitute the necessary "blueprint" for the actual installation program.

The steps involved in implementing this blueprint are (1) apply for necessary permits in a timely manner, (2) write detailed specifications for all materials and equipment, (3) determine which materials will be ordered by the owner and which by contractors, (4) order materials to be obtained by the owner in a timely fashion, (5) write a request for bids for construction contracts, including the specifications for the materials to be supplied by the contractors (all bid requests should include deadlines to be met and performance guarantees required), (6) award the purchase and construction contracts, (7) supervise and/or assist during construction phase to ensure that permit requirements, specifications, schedule, and budgets are met, (8) train personnel, and (9) start up facilities.

Some steps on this list are discussed in more detail later in this chapter. All are important to ensure a successful waterflood installation. For smaller projects, the waterflood engineer will be deeply involved and may be required to do all these steps in person. For larger projects or companies, only peripheral involvement of the waterflood engineer in some or all these steps may be required. However, the engineer should remain as informed and involved as possible to ensure that the design is implemented and will perform as planned.

12.3 Responsibilities

As mentioned in Sec. 12.1, the engineer's responsibility will vary, depending mainly on project and company size. Another factor affecting his involvement is the type of construction contract chosen. Most waterflood agreements will be a partial or complete "lump-sum" (turnkey) type or a "time-and-materials" contract. In the former, the contractor is fully responsible for the work detailed in the request for bid. Any changes or additions to the scope of the work will generally be very expensive and should be made only if absolutely required. Normally, the engineer will only monitor the work done under a full turnkey contract to ensure compliance with the design and specifications. In contrast, a time-and-materials contract will necessitate very close supervision by the engineer or other owner-company personnel. In this type of agreement, the unit cost (costs per hour, per foot of pipe, per cubic yard of concrete, etc.) will be specified in the bid, but the hours to be worked and the work to be done are not included. A time-and-materials contract often requires that the work be directed by representatives of the owner. The engineer will likely be heavily involved technically and may be needed for on-site supervision.

The design engineer may have legal responsibilities. This liability can have at least two aspects that can lead to lawsuits against the engineer: inherent design flaws and direct supervision orders. Most large, and many smaller, companies indemnify their engineers against such lawsuits. The engineer should consult legal counsel (private or company) concerning personal responsibilities, exposure to legal liability, and the company's policies.

Overall, the engineer's responsibility is to ensure that the waterflood is built as designed, on time, and as budgeted, with such modifications as are deemed necessary and justified. The actual duties of the engineer will vary, depending on the project, but the goal remains the same.

12.4 Specifications

Two major areas require specifications to install a waterflood successfully: materials or procurement processes and labor or construction efforts. Purchasing materials for the project requires specifications that are detailed, technical descriptions based on known engineering standards or references. Specifications for construction work focus on a description of the job to be done (possibly in detail) and the legal aspects of the contracts between parties. This involves a clear understanding of the responsibilities and relationships between the parties. Arnold and Stewart[7,8] list the following items for inclusion in specifications: (1) bid scope, (2) contractor's responsibility and scope of work, (3) design conditions (pressure, temperature, fluids analyses, etc.), (4) company-furnished materials, (5) bid details, (6) timing requirements (deadlines), (7) inspection and testing by owner and contractor, (8) acceptance conditions, (9) preparation for shipment and shipping method, (10) warranty with terms required, (11) cost and method for change orders or extra work, (12) payment schedules, (13) contractor-furnished data, and (14) technical specifications, including materials, reference standards, and procedures. All the items on this list may need to be considered when specifications are developed for either materials or construction.

12.4.1 Materials. The specifications for materials must be carefully written to ensure that the correct items are purchased. Normally, engineering standards are referenced in such areas as design criteria, pressure/temperature ratings, welding techniques, metallurgy, and testing/inspection methods. In some standards, a vendor data sheet is included. Examples of this practice may be found in API Standard 610, Appendix A,[9] and API Standard 674, Appendix A.[10] A typical data sheet for pumps includes sections on liquid properties, operating conditions, site conditions, performance requirements, construction details, materials, auxiliary piping, inspections and tests, prime mover, weights, and standards to be satisfied. Many companies have specifications already written for the more common items ordered. The engineer should use these specifications, if available and applicable. Typically, such prewritten documents will require only that blank spaces be filled in for the particular item needed.

12.4.2 Construction Labor. Depending on the type of contract (Sec. 12.6), the specifications for labor may be very general or extremely detailed. The requirements for turnkey contracts must be covered in depth, including referenced engineering standards, quality-assurance testing procedures, and most of the 14 items listed previously. Such details as burial depth for pipelines, qualification testing for welders, radiographic testing procedures (percent of welds to be examined and by whom), welding procedures allowed, curing procedures for fiberglass pipe bonds, wiring-harness design, and Natl. Electrical Manufacturers Assn. (NEMA) and American Natl. Standards Inst. (ANSI) standards are only a few of the possible items to be included in the specifications. Responsibility for performing and paying for testing of welds and repairing failed joints should be specified. This level of detail requires substantially more engineering time to ensure that the design is complete before bid requests are written. The waterflood engineer should consult appropriate experts to help resolve these design specifications as necessary.

In contrast, a time-and-materials agreement may specify only the general scope of work. The contract in this case may contain only a labor rate schedule and an equipment-rental rate schedule. Labor disciplines might include such personnel as welders, electricians, heavy-equipment operators, foremen, general laborers, painters, and other needed skills. Equipment could include welding machines, ditchers, bulldozers, cranes, cement mixers, side booms, and various kinds of trucks. These listings are far from complete, but give a general idea of the items in time-and-materials contracts.

Each waterflood will require labor specifications that are tailored to the particular needs of the project. The engineer must design the specifications to fit the waterflood needs, depending on contract type, regulatory and physical environment, scope of the project, and available resources.

12.5 Purchasing and Logistics

Procurement and timely delivery of equipment and materials is critical to finishing a project on time and on budget. As discussed earlier, this task usually is performed by a separate purchasing group

Requirements	Companies Bidding		
	A	B	C
Cost			
Base			
Tax			
Freight			
Total			
Size/model			
Metallurgy			
Pressure rating			
Temperature rating			
Design flow rate			
Maximum flow rate			
Flow rate at maximum pressure			
Minimum flow rate			
Horsepower required			
Fuel/power usage			
Weight/dimensions			
Delivery time			
Comments			

TABLE 12.1—BID EVALUATION FORM

for major projects or in large companies. For smaller projects or companies, the waterflood engineer may be responsible, directly or indirectly, for this function.

In either case, the first step is to order the items with long delivery times identified in the design process. Major items such as pumps, prime movers, vessels, and treating systems usually will be in this category. Major supervisory-control or electrical systems may require long lead times. Any items requiring special metallurgy or elastomers for construction also may need extra time. Standard tanks, line pipe, fittings, valves, instruments, controls, and electrical items are generally available off the shelf and hence have short delivery periods. To minimize costly delays, the engineer should use standard parts as much as possible, commensurate with project requirements—e.g., corrosion protection should not be sacrificed.

Specific major items should be purchased on a bid basis to ensure that the lowest effective price is paid. Other materials required in quantity (line pipe, fittings, and cement) can be bid on a unit-price basis. Some items may be included in a subcontractor's or vendor's bid for fabrication or manufacture of equipment, particularly if skid-mounted or constituting a separable subsystem. Obtaining at least three bids (five or more are not uncommon) by qualified vendors for each bid request is recommended. Qualifications must include financial soundness as well as technical abilities. In-house business personnel should be able to evaluate the former for each bidder. Bid evaluation requires the engineer to ensure that different bids are compared on an "apples-to-apples" basis; separating each bid into common pieces or features will allow an accurate comparison. Table 12.1 shows an example form for this type of analysis. Although a number of formats are possible to meet specific needs, the key element is to compare the bids on the basis of the separate requirements of the specifications. The bid should be awarded to the bidder who is most cost-effective—i.e., who has the best combination of initial price, operating costs, and performance. Transportation should be included to compare costs at the worksite.

The engineer should follow the policies of the owner company concerning bid requests, openings, evaluation, and awarding. If no policies have been set, the previous discussion contains guidelines for bid requests.

Bid openings are another area of concern. If sealed bids are requested, the engineer should examine all bid envelopes (before opening) for evidence of tampering or lack of sealing. If such evidence is found, a good practice is to return that bid to the vendor for rebidding without opening any bids. A new time for bid opening should then be set. Each bid should be opened in the presence of at least one or two additional company personnel. Some companies require proof of this in the form of the witnesses' initials on the bids. All bids should be opened during the same time period. It is considered unethical[11] to disclose any vendor's bid to personnel from any company other than the owner's. Bid shopping by using one bidder against another is discouraged. As Arnold and Stewart[7] write: "...this will be counterproductive. If each vendor expects the work to be given to the low bidder, he will put his best effort into the price. If an auction is expected, the bid price will reflect this." The engineer is advised to keep this statement in mind throughout the bidding process.

The next step in the procurement process is to award the contract, which may be in the form of a purchase order or some other company-specified format (contracts or term agreements). The engineer should be sure that proper documentation and approvals are obtained.

"Logistics" may be defined as the science of obtaining and delivering the proper quantity of needed items to the place that needs them in a timely manner. Obtaining the needed items has already been discussed. In large projects, a staff of expediters may be available to ensure delivery to the right place at the right time. In smaller projects, the engineer may perform this function. The critical items usually are those with long deliveries. Other factors that increase logistical difficulties are remote locations, such as offshore, arctic, mountainous, or marshy terrain, and areas not currently served by a petroleum industry infrastructure. Where these conditions exist, logistical control becomes crucial, and the engineer may have to devote extensive efforts to expediting materials.

12.6 Contractor Relationships

Construction contracts can fall into four different formats[8]: turnkey, negotiated turnkey, modified turnkey, and cost-plus contracts. A time-and-materials contract is a form of modified turnkey agreement. Frankhouser[12] categorizes turnkey, time-and-materials, and cost-plus as lump-sum, unit-price, and reimbursable contracts, respectively. These contract types decrease the predictability of the cost and increase owner risk, control, and involvement in the order presented. There are advantages to each type of format, as detailed in Refs. 11 and 12.

The turnkey contract places all responsibility and risk on the contractor, but the owner loses control over construction. The cost is fixed, but may be higher than other formats as a result of the increased risk for the contractor. Turnkey projects also require a very high level of detail in describing the scope of the work. Because additions or changes to the work detailed in the bid will require new negotiations with the contractor and only the turnkey contractor is likely to be able to do the work economically, such changes are frequently very expensive. The engineer should make every effort to eliminate or to minimize these "extras."

A negotiated turnkey is not pertinent to this discussion, but involves use of a contractor for the engineering design (for a fixed fee) and then negotiation of a turnkey price with that contractor for the actual installation.

In the modified turnkey, the owner company bids out the materials and labor for each major work segment and then coordinates and expedites these jobs. The owner thus has more control, but must assume more risk and contribute more manpower to the project.

A cost-plus agreement, normally used where uncertainties in the scope and risk are high, is rare in waterflood projects. The contractor is paid all actual costs plus a fixed fee or, in some cases, a fixed percentage for overhead and profit. The owner has substantial control over the contractor's work but bears all the risk in the fixed-percentage case. With the fixed-fee arrangement, the contractor bears some of the risk. When the agreement is signed, the owner will have a high degree of uncertainty about the total costs involved.

Turnkey contracts often require more time to complete an installation because of additional time needed either to prepare the detailed specifications and job description or, in modified turnkey contracts, to bid each segment separately. The cost savings realized may not justify the increased delay.

As a general rule (although exceptions abound), a waterflood installation contract will be either a modified turnkey or a time-and-materials agreement. In two comparisons cited by Arnold and Stewart,[8] the modified turnkey approach costs 10 to 13% less than negotiated turnkey and turnkey projects. This approach involves dividing the project into separate segments, each of which is bid

out on a turnkey basis. Owner-company personnel then coordinate the various segments to meet schedules and budgets. For relatively small projects, an experienced waterflood engineer or construction superintendent can frequently minimize the cost under a time-and-materials agreement. As a point of caution, the experience level of the owner's construction supervisor is crucial in achieving minimum expense. For the inexperienced waterflood engineer, hiring an experienced construction firm under a modified or even a complete turnkey contract may be advisable. Another approach in this case is to hire an experienced consultant to do the on-site supervision.

The waterflood engineer must evaluate the nature of the project, including scope, size, complexity, and cost, compared to the resources available, including in-house experts, consultants, contracting firms, and vendors. This appraisal must be very objective and include the abilities of the engineer. This will enable the optimum choice to be made concerning specifications and contracting for both purchasing and construction activities.

12.7 Quality Assurance

Throughout the construction process, quality assurance is very important to ensure that the facility is built as designed, to specification, on time, and within budget. ANSI[13] defines quality assurance as "all those planned and systematic actions necessary to provide adequate confidence that an item or a facility will perform satisfactorily in service." These actions should include the necessary controls and administrative functions to ensure that the money spent on the project is accounted for and expended in an effective manner. The engineer should work with financial and administrative personnel to maintain full compliance with company policies and standards.

Quality control is a part of quality assurance that focuses on the equipment, materials, and construction procedures. For larger projects, a team of inspectors may be available to fulfill this function. In smaller waterfloods, the engineer may perform these duties directly. The specific actions carried out include inspection and testing to ensure that specifications and standards are satisfied by the vendors and contractors. The engineer or inspector may need to make several trips to a vendor's facility to observe the fabrication process, including all tests required by the specifications. These tests can include radiographic and metallurgical techniques as well as physical examinations for size, quality of finish, film thickness, or other similar factors. Hydrotesting of injection lines, piping systems, and vessels is another common testing procedure needed. The engineer should maintain complete familiarity with the standards referred to in the specifications, including the required testing methods. Vendors and contractors should abide by all such specifications. Inspection and testing (or witnessing these activities) during the procurement and fabrication process will ensure compliance.

In many cases, availability of personnel may preclude adequate on-site trips to inspect all equipment and materials. When this occurs, the engineer must ensure that the high-priority items are sufficiently checked for safety and performance. Another protective method under these conditions is to examine the qualifications of the bidders carefully. The most reliable and qualified bidder should be chosen, even if some additional cost is involved. The expense of more personnel can be balanced against the higher vendor price (sometimes) needed for adequate reliability. Wood[14] calls this an appraisal of potential vendors' quality systems. In other words, if the engineer cannot ensure complete inspection and testing, then the contractor must have systems in place that guarantee the satisfactory construction and performance of the equipment.

As a general rule, any reasonable quality-assurance effort will be justified economically by the longer-term reliability and performance of the equipment. The engineer should ensure that proper procedures are followed and that enough personnel are assigned to this function. Only if this is done will the facility successfully perform in service and be the most cost-effective.

12.8 Startup

For major waterfloods, such as Prudhoe Bay in Alaska, a complete startup team is formed to initiate and "debug" the operation of the facilities. At Prudhoe Bay, some members of this team were retained as part of the operational personnel, but others returned to other jobs or projects. Fraylick[15] discusses many aspects of a major project startup partially based on the Trans-Atlantic Pipeline System experience. For the purposes of this section, "startup" includes precommissioning efforts in addition to actual startup of the entire project.

For the smaller projects concentrated on in this chapter, a full-time startup team is not feasible. The startup group will usually comprise the design engineer(s), construction supervisor(s), manufacturer's representative(s), and operational personnel. Before startup, the operational personnel must be adequately trained. As mentioned in earlier chapters, proper documentation of the equipment and facilities must be available. At the least, this should include operation and maintenance manuals, as-built drawings, and specifications. Such documents greatly facilitate the training process. If the operations personnel have been consulted in detail throughout the design phase (as advocated in this monograph), the need for training will be lessened but not eliminated. Both classroom and hands-on experience should be given to operating personnel so that they understand normal operation and what can go wrong. These personnel can then be used as an integral part of the startup process and move smoothly to the operational phase.

Startup procedures are specific to a given project, but a few guidelines common to most cases can be given. A schedule and a written plan should be developed for all startups. Careful, detailed planning at a level appropriate to the project's size and complexity will pay off in an easier, faster, and more successful startup process. The effort needed can vary from a few to many thousands of work-hours. For example, the production facilities for the Beta field offshore California required about 30,000 work-hours[16] for startup, of which a substantial amount went into planning. The engineer must determine what effort is appropriate for planning and scheduling and ensure that it is expended.

Another procedure commonly followed during precommissioning and startup is to make a list of all items that did not work properly and require fixing or that were not finished. This is often called a "punch list" or "but list" and serves as written documentation of the problems that need correction. Whenever possible, the contractor's representatives should be shown the problem and concur with the need for repairs. Determining which party will pay for the repair or completion of an item may present a problem, but can be negotiated. If possible, the original contract should provide for the development of this final punch list and set guidelines for determining responsibility. This procedure should be part of an overall plan for turning over the facility after acceptance by the operating supervision.

The planning for startup should include personnel requirements, particularly in such specialty areas as instrument technicians, vendor representatives, electricians, and computer technicians. Any changes made during startup should be temporarily "red-lined" on the working drawings for later incorporation into the as-built drawing set. Individual pieces of equipment may be started up for testing, and then shut down again until actual startup of the entire system. The engineer should evaluate the utility of "dry runs" for individual pieces of equipment or entire systems. Piping systems and vessels should be flushed out and all extraneous materials removed. The alignment of pumps and prime movers should be verified for tolerances. All lubricants should be installed or, if factory-installed, checked for level and quality. Cooling systems should be flushed as appropriate, and the proper amount and type of coolant installed or verified. Instruments and meters must be calibrated and their actions or signals be checked against specifications. Instrument and fuel gases need to be sampled to ensure that the water content, heating value, oxygen, and inert contents are within the correct limits. Sight gauges or other visual instru-

ments should be cleaned for proper operation. Sample valves and lines need to be flushed. All necessary chemicals should be on hand and connected to the pumps or other means of delivery. The chemical systems will also need to be cleaned and flushed. Whenever it is possible and safe to do so, all electrical circuits should be energized in dry runs. These suggestions include some of the normal items associated with startup of a waterflood project. The engineer is urged to tailor startup planning carefully to fit the individual project. Only then will the actual startup have a reasonable chance of proceeding satisfactorily.

After startup, a review of the actual events compared with the planned activities is usually made. A report documenting significant variances and discussing the probable cause(s) can help prevent errors. Favorable differences should be included as well. A reconciliation of variances with causes can lead to increased efficiency and success on future project startups and serves as an excellent teaching tool. Following startup, the red-lined working drawings should be incorporated into a final set of reproducible as-built drawings. These, plus the operations and maintenance manuals, vendor literature, specifications, and other appropriate documents, should be turned over to the operations personnel. Because these drawings, manuals, and documents are very important in the proper long-term operation of the waterflood system, it is recommended that the contract, purchase order, or requisition stipulate that final payment will not be made until this information has been received. This will help ensure the smooth, successful functioning of the waterflood project.

Throughout the process of designing and installing a waterflood, common sense and technical ability, combined with a willingness to learn from operations and other personnel, will enable the engineer to be successful. This accomplishment will create job satisfaction and enrichment for the individual and a good economic investment for the company. It is hoped that this monograph will contribute to the success of such projects. Good luck!

References

1. Gregg, D.E.: "North Sea Project Management," *JPT* (Nov. 1983) 1941–49.
2. Hill, V.F.: "Project Poisons and Antidotes," *JPT* (Feb. 1984) 281–86.
3. Clawson, R.M. and Cloughley, T.M.G.: "A System for Project Planning and Control," paper EUR 199 presented at the SPE 1980 European Offshore Petroleum Conference, London, Oct. 21-24.
4. Shaffer, L.R., Ritter, J.B., and Meyer, W.L.: *Critical Path Method*, McGraw-Hill Book Co., New York City (1965).
5. Lockyer, K.G.: *An Introduction to Critical Path Methods*, Pitman Publishing Corp., New York City (1964).
6. Schoemaker, R.P.: "Visual Path Analysis," *JPT* (Feb. 1967) 177–85.
7. Arnold, K.E. and Stewart, M.: "How to Manage Production Facility Projects," *World Oil* (Nov. 1983) 62–66.
8. Arnold, K.E. and Stewart, M.: "How to Manage Production Facility Projects," *World Oil* (Dec. 1983) 71–76.
9. *Standard 610, Centrifugal Pumps for General Refinery Service*, sixth edition, API, Dallas, TX (1981).
10. *Standard 674, Positive Displacement Pumps—Reciprocating*, first edition, API, Dallas, TX (May 1980).
11. Vicklund, C.A. and Craft, W.S.: "Management of Major Offshore Projects—An Industry Challenge," *JPT* (April 1981) 585–93.
12. Frankhouser, H.S.: "Project Management (North Sea)—Basic and Variable Factors," *JPT* (Oct. 1981) 1821–27.
13. "Quality Assurance Terms and Definitions," American Natl. Standards Inst., New York City, N-45.2.10 (1973).
14. Wood, K.: "Quality Assurance Programmes for Offshore Production Platforms," paper EUR 202 presented at the 1980 SPE European Offshore Petroleum Conference, London, Oct. 21-24.
15. Fraylick, J.R.: "How to Start Up Major Facilities, Parts 1 and 2," *Oil & Gas J.* (Dec. 3, 1984) 112–15; (Dec. 10, 1984) 92–94.
16. Visser, R.C.: "The Beta Field Development Plan," *JPT* (Nov. 1982) 2517–22.

Appendix A
Major Project Execution Procedures

Definitions and Outline

Typical steps for project execution are outlined in Table A-1. Listed with each step are the personnel with primary responsibility. Also listed are other personnel who may have pertinent input for each specific step. Part of the primary responsibility is to ensure that adequate input is obtained from other groups when needed. The following personnel are involved.

The *coordinator*, representing the local organization's management in the project's execution, coordinates the liaison with and control over all outside (if any) organizations, such as the purchasing department, centralized project engineering groups, and consultants.

The *project engineer* is in overall charge of the engineering, construction efforts (usually), and personnel on the project. In many cases, the project engineer will also be the coordinator when the authority for the project lies within one organization.

Production includes the personnel and managers directly responsible for production operations in the field. They may cover only the field in which the project is proposed or some larger area, but are part of the local organization—e.g., district, area, or division.

Reservoir includes the personnel and managers directly responsible for reservoir analysis and engineering within the local organization.

The *staff* includes the personnel and managers who are not directly responsible for certain areas but rather are expert resources in various technical fields. They may be part of the local organization or a centralized headquarters.

Purchasing includes the personnel and managers directly responsible for purchasing equipment and materials, usually a centralized group.

Materials includes the personnel and managers directly responsible for the receipt, control, and disposal of equipment and materials within the local organization.

Specific Review and Sign-Off Procedures

To ensure that the required personnel have provided input to the various project phases, the coordinator will obtain and maintain a list of approval signatures from the groups listed above. Appropriate signatures will be obtained for each major project step. Areas of particular significance are discussed below.

Project Scope Definition and Acceptance. After the best alternative solution is determined, the specific project scope of work will be formulated and agreed upon by the coordinator, production personnel, and project engineer. The local engineering and operations managers will sign off on the agreed scope. This scope forms the authorization for expenditure (AFE) budget basis, so such concurrence is required to avoid "downstream" scope changes and resultant budget additions.

To ensure a comprehensive scope of work, a meeting of appropriate personnel from these three groups will be held before the AFE cost estimate. It is anticipated that several work items will result from this meeting to ensure that the scope encompasses all known concerns. After all concerns are addressed, a second meeting will be held to achieve project scope concurrence.

Development/Review/Acceptance of the AFE Cost and Schedule Estimates. Once local and headquarters concurrence to the project scope is achieved, the project engineer will develop AFE quality cost and schedule estimates. The coordinator will review the estimate, comparing the estimate to similar projects, and/or comparing project components to similar components of different projects. Once the coordinator accepts the estimate's validity, he will issue a letter stating this and comparing the estimate to known costs of similar projects/components. This letter will be issued to local engineering and operations management.

Development of Process Flow Sheet and Piping and Instrumentation Diagrams (P&ID's). These drawings form the basis for all project execution tasks. In the AFE schedule estimate, the project engineer will commit to a distinct development duration for these drawings. The project engineer will issue biweekly status reports to the coordinator during this phase of development. Once developed, the P&ID's will be reviewed by the coordinator and production personnel for scope adherence. The P&ID's will be completed when the coordinator, project engineer, and production management agree to their content by signing off on the drawings.

Detailed Engineering Design. After the P&ID's are completed, the detail design can be performed. This step must be executed completely to ensure comprehensive and efficient material procurement and construction project phases. The detail design disciplines include process, piping, mechanical, instrumentation, structural, and electrical. The project engineer will schedule regular meetings during this phase to inform the coordinator of the design progress. Other local and/or headquarters personnel may attend these meetings as appropriate.

Material Procurement. Material delivery is frequently the project schedule bottleneck. Major items (and items with long deliveries) can be ordered before design completion; however, smaller items will be ordered upon a material take-off after completion of the detailed design. The coordinator will inform the project engineer of any local surplus material to be used on the project. The project engineer will coordinate between purchasing and transportation personnel to ensure that the lowest purchase bid is still the lowest with freight included. During selection of new equipment, the project engineer, in consultation with the coordinator, will consider its compatability with any existing facilities to take advantage of spare-parts inventory. The operating personnel's familiarity with existing facilities and their developed competence in the operation of such equipment is a positive element that should be used where practical.

As early as possible in the project, purchasing should be consulted to identify long lead items. This consultation should take place before completion of detail design.

Materials personnel are responsible for proper storage and maintenance of all equipment after delivery but before installation. This time period should be minimized as much as practical to prevent loss or damage.

TABLE A-1—PROJECT EXECUTION STEPS

Project Step	Primary Responsibility	Other Sources of Information When Pertinent
Conceptual Design		
Identify problems, list alternative solutions	Coordinator, production	Reservoir, local staff, headquarters staff
Make preliminary cost and schedule estimates and alternatives, evaluation/ recommendation	Coordinator	Production, reservoir, local staff, headquarters staff
Define project scope and set design conditions/ requirements*	Coordinator, production, project engineer	Reservoir, local staff
Obtain required permits	Environmental group, production, coordinator	Local staff, headquarters staff, reservoir
Develop process flow sheet	Project engineer, coordinator	Production, headquarters staff
Develop P&ID's	Project engineer	Coordinator, production, headquarters staff
Project AFE		
Develop AFE cost and schedule estimates	Project engineer, coordinator	Headquarters staff
Develop AFE package	Coordinator	Local staff
Solicit AFE approvals	Coordinator	Local staff
Detail Design		
Provide status reports (monthly except as noted otherwise)	Project engineer, coordinator	Headquarters staff, local staff
Develop detailed drawings and specifications	Project engineer	Coordinator, production, headquarters staff
Develop material take-off	Project engineer	Coordinator, production, headquarters staff
Procurement		
Purchase material (considering transportation costs)	Project engineer, purchasing	Coordinator, materials
Expedite material	Project engineer, purchasing	Coordinator
Inspect fabrication	Project engineer	Coordinator
Dispose of surplus material	Project engineer	Coordinator, production, materials
Construction/Installation		
Bid/evaluate/award construction contract	Project engineer	Coordinator, production
Provide comprehensive inspection	Project engineer	Coordinator, production
Develop and complete punch list	Project engineer, coordinator, production	
Provide as-built drawings and data books	Project engineer	Coordinator
Check-Out/Startup		
Start up AFE	Coordinator, production	Project engineer
Prepare operating/training procedures, conduct functional check-out, start up facility	Production, coordinator	Project engineer, headquarters staff, local staff
Provide project reconciliation and recommendations	Coordinator, project engineer	Production, headquarters staff, local staff

*Because of the importance of scope definition, the local engineering and operations managers should sign off on the agreed scope.

Material purchased by contractors must comply with company material standards.

The project engineer should ensure that vendors supply appropriate installation, operation, and maintenance manuals. Drawings showing the external and internal details of the equipment should also be obtained. The project engineer should include this material in the data books for the project.

Surplus project material remaining at the end of the project should be disposed of by the project engineer. Production should be given a chance to retain any useful material; however, it remains the responsibility of the project engineer to dispose of the surplus and to credit the AFE accordingly.

Construction Contract. The project engineer will determine the most cost-effective contracting strategy—i.e., lump-sum or time-and-material contracts. This evaluation is unique to every project. The local management will review the contracting strategy and the potential contractors for the job before bid solicitation. The project engineer will issue a construction schedule and budget to the coordinator at construction mobilization and will update it biweekly during this phase.

Inspection Services. The project engineer will ensure inspection of major purchased items and of construction procedures. Local management will reserve the right to perform an acceptance inspection on equipment before it leaves the fabrication shop (timely notification from the project engineer is essential for this effort). This acceptance inspection will provide an operational inspection viewpoint, may catch previously undetected errors, and will facilitate the on-site inspection after delivery.

Inspection of contractors by contracted inspectors has frequently resulted in problems. The local management can request that company inspectors be used. All contractors are required to comply with safety and environmental rules and regulations.

Punch List. Project engineer and local management concurrence to a "punch list" is required to define contractor demobilization.

A punch list is a tabulation of all items that have not been completed or that require corrections or repairs. This list is used to ensure that the project is properly completed before the contractor is released. All uncompleted items in the project scope of work that are discovered in the check-out should be completed or added to the punch list. Completion of the punch list ends the construction phase and usually initiates the startup phase. The local engineering and production management will sign off on the agreed-to punch list.

As-Built Drawings and Data Books. The project engineer will issue as-built drawings and data books to the coordinator within 4 weeks of punch-list completion. Vendor data required for startup will be provided before that time. The project engineer will expedite these data to meet this time requirement.

Startup AFE. The coordinator will prepare and submit for timely approval the check-out/startup AFE. Direct input from production personnel is required to estimate costs and timing.

Project Reconciliation. A complete cost and schedule performance reconciliation should be performed by the project engineer and coordinator to determine areas of improvement for future projects. This reconciliation, comparing costs against budget estimates, should be issued in a letter to local and headquarters management within 2 months of the project startup.

The main project responsibilities lie between the project engineer and coordinator. Careful, upfront planning and strict adherence to the proper sequence of project phases should enable successful performance within the budgeted estimates. Realistic budget and schedule estimates must be accepted by management. Travel expenses to maintain effective communication through progress meetings should be anticipated.

Acknowledgment

This Appendix is based on an unpublished technical paper by D.G. Siekkinen, Arco Oil and Gas Co., July 1985.

Appendix B
Example Calculations

Example Calculation for Pipe Support Spacing

It is given that two quintuplex pumps are manifolded together, 250 strokes/min, with 4-in., Schedule 40 line pipe on the discharge. Calculating the highest estimated pump-pulsation frequency, we get

$$F_p = \frac{250 \times 5 \times 2}{60} = 41.667 \text{ pulses/sec}$$

and

$$f_p = 4 \times F_p = 166.67.$$

Now, calculating support spacing from Eq. 6.2, we obtain

$$L = \left[\frac{1.69}{166.67} \times \left(\frac{30 \times 10^6 \times 7.23}{10.79} \right)^{0.5} \right]^{0.5}$$

$$= 6.75 \text{ ft.}$$

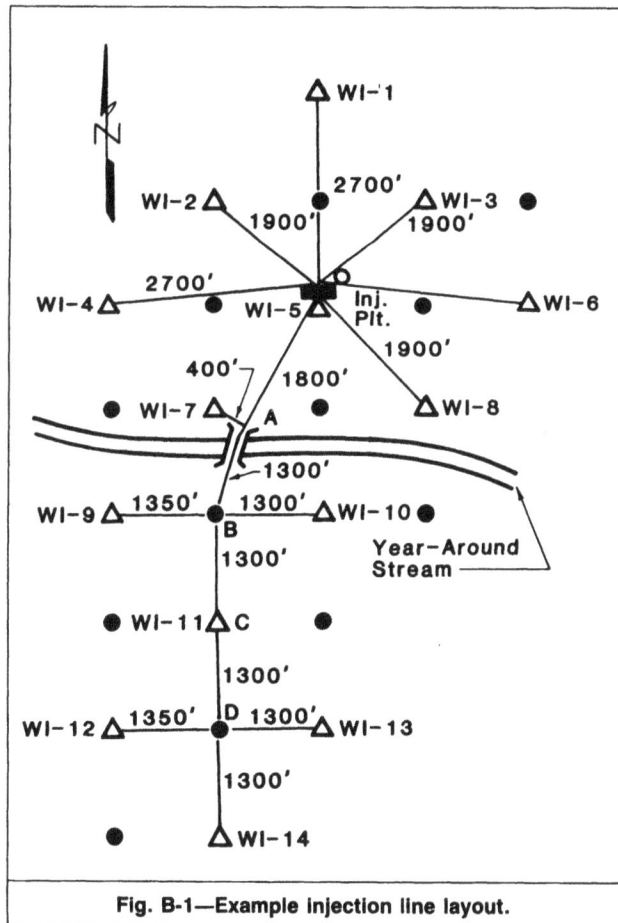

Fig. B-1—Example injection line layout.

Example Calculation for Line Sizing

See Fig. B-1 for the schematic layout of the system. With the Darcy formula (Eq. 6.3) and Moody friction factors, the pressure drops by line segment can be calculated for the trunk/lateral system. The Moody friction factors are available in the *Trans.*, American Soc. of Mechanical Engineers (1944) **66**, 671.

Case 1. Assume that Trunkline OD is 2-in. nominal, Schedule 40 steel line pipe and all laterals are 1½-in. nominal, Schedule 40. The Reynold's number is

$$N_{Re} = 50.6q \times \rho_w/(d \times \mu_w),$$

where

q = flow rate, gal/min,
ρ_w = water density, lbm/ft³,
μ_w = water viscosity, cp, and
d = ID, in.

The values for Line Segments OA, AB, BC, and CD are given in Table B-1.

Injection rates for wells off the trunkline are 7.3 gal/min for Wells WI-7, WI-10, and WI-11; 5.8 gal/min for Well WI-9; 4.4 gal/min for Wells WI-12 and WI-13; and 4.3 gal/min for Well WI-14. Then, by inspection, the lateral with the maximum Δp should be to Well WI-10 because it has the same injection rate and a longer flowline than Well WI-7 or WI-11. From B to Well WI-10, the injection rate is 7.3 gal/min, $N_{Re} = 22{,}000$, $f = 0.0286$, $\Delta p = 2.5$, and cumulative $\Delta p = 42.7$. The allowable pressure drop is 10% of the maximum system pressure, or 100 psi, in this case. Therefore, the pipe sizes should be reduced.

Case 2. Assume that Segment OA of the trunklines is 2 in., the remaining trunklines 1½ in., and the laterals 1¼ in. Then, with the flow rates remaining as shown in Case 1, Table B-2 gives parameter values for Line Segments OA, AB, BC, and CD, and from B to Well WI-10. Table B-2 shows that further reduction is possible.

Case 3. Keep the same assumptions as Case 2, except reduce the laterals to 1 in. For Line Segment OA to CD, $\Delta p = 57.5$ and cumulative $\Delta p = 57.5$. From B to Well WI-10, $N_{Re} = 33{,}000$, $f_p = 0.0273$, $\Delta p = 20.5$, and cumulative $\Delta p = 78.0$. This shows that still further reduction is possible.

TABLE B-1—PARAMETER VALUES FOR CASE 1 LINE-SIZING CALCULATION					
Segment	q	N_{Re}	f	Δp (psi)	Cumulative Δp (psi)
OA	40.8	91,000	0.0221	22.6	22.6
AB	33.5	74,000	0.0228	11.2	33.8
BC	20.4	44,000	0.0242	4.3	38.1
CD	13.1	30,000	0.0255	2.1	40.2

TABLE B-2—PARAMETER VALUES FOR CASE 2 LINE-SIZING CALCULATION

Segment	N_{Re}	f	Δp (psi)	Cumulative Δp (psi)
OA	91,000	0.0221	22.6	22.6
AB	95,000	0.023	22.2	44.8
BC	56,000	0.0245	8.6	53.4
CD	39,000	0.0255	4.1	57.5
B to Well WI-10	25,000	0.028	5.3	62.8

TABLE B-3—PARAMETER VALUES FOR CASE 4 LINE-SIZING CALCULATION

Segment	N_{Re}	f	Δp (psi)	Cumulative Δp (psi)
OA	91,000	0.0221	22.6	22.6
AB	95,000	0.023	22.2	44.8
BC	65,000	0.0248	33.2	78.0
CD	45,000	0.0258	15.9	93.9
B to Well WI-10	33,000	0.0273	20.5	114.4

TABLE B-4—PRESSURE DROPS FOR CASE 5

Segment	Δp (psi)	Cumulative Δp (psi)
OA	22.6	22.6
AB	22.6	44.8
BC	8.6	53.4
CD	15.9	69.3
B to Well WI-10	20.5	89.8

TABLE B-5—VELOCITIES FOR CASE 5

Line Size (in.)	Maximum Flow Rate (BWPD)	Maximum Velocity (ft/sec)
2	1,350	3.8
1½	1,100	6.8
1¼	450	3.9
1	250	2.7

Case 4. Segment OA is 2 in., AB is 1½ in., BD is 1¼ in., and laterals are 1 in. The values of N_{Re}, f_p, Δp, and cumulative Δp are shown in Table B-3. The pressure drop from B to Well WI-10 is over the allowable amount, so line size must increase.

Case 5. Final sizing is Line Segment OA at 2 in., Line Segment AC at 1½ in., Line Segment CD at 1¼ in., and the laterals at 1 in. Values for Δp and cumulative Δp are shown in Table B-4. This set of line sizes must now be checked to ensure that the velocities are acceptable. Eq. 6.6 calculations give the values in Table B-5. These values are within acceptable tolerances, so this is the final sizing solution.

In the northern section of the waterflood, a radial system appears to be most practical. By inspection, it appears that the line to Well WI-6 should have the greatest pressure drop. So if this segment has a $\Delta p < 100$ psi for a given line size, then all the other radials will also be below that limit. For 1-in. pipe, the value of N_{Re} is 26,500, $f=0.029$, and $\Delta p=14.5$ psi. For ¾-in. pipe, we get a Δp of 48.4 psi and for ½-in. pipe, $\Delta p=204$ psi. Hence, the proper line size for the radial injection lines is ¾-in. nominal based on these calculations.

In normal oilfield practice, ¾- or 1-in. nominal pipe is too small to install and service for runs of 1,000 ft or more. A line of this size is susceptible to damage, particularly if laid on the surface.

Therefore, it is likely in the case above that 1½- to 2-in. nominal pipe would be used throughout the system.

Nomenclature

d = ID, in. [cm]
f = Moody friction factor, dimensionless
$f_p = 4 \times F_p$
F_p = highest pulsation frequency, pulses/sec
L = length between supports, ft [m]
N_{Re} = Reynolds number
Δp = pressure drop, psi [kPa]
q = flow rate, gal/min [m³/min]
μ_w = water viscosity, cp [Pa·s]
ρ_w = water density, lbm/ft³ [kg/m³]

SI Metric Conversion Factors

bbl	× 1.589 873	E−01	= m³
ft	× 3.048*	E−01	= m
gal	× 3.785 412	E−03	= m³
in.	× 2.54*	E+00	= cm
psi	× 6.894 757	E+00	= kPa

*Conversion factor is exact.

Appendix C
Definitions of Economic Parameters

Cash Flow

Cash flow can be defined as the difference between revenues received and cash disbursed over a given time period. All cash exchanges should be included.

Present-Worth Values

The present-worth concept is an approach to evaluate the time value of money. It does so by "discounting" the value of future dollars. Discounting can be defined as the process of calculating the current (or present) value of a future dollar received or spent. For example, 32¢ invested today at 12% interest (compounded continuously) will yield $1.00 in 10 years. Conversely, then $1.00 to be received 10 years from now is worth 32¢ currently. Hence, the discount factor for 12% rate, compounded continuously for 10 years, is 0.32. The present-worth value at a given discount factor is the sum of the discounted cash flows.

An example of this calculation is shown in Table C-1 for a continuously compounded 12% rate for a project starting Jan. 1, 1988, and ending Dec. 31, 1992. The present-worth value discounted at 12% for this project is $1,560. Discount factor tabulations are widely available for different interest rates and different compounding techniques—e.g., uniformly (continuously) or instantaneously (at discrete points in time) over various time periods.

Investment Efficiency

Investment efficiency is the measure of the present worth per dollar invested. It may also be called the present-worth index or profitability index. This parameter is most commonly used for capital allocation among alternate projects. It is calculated by dividing the present-worth value at a given rate by the investment. For the example in Table C-1, this index equals 1,560 divided by 8,000, or 0.2. Generally, values must be in the 0.5 to 0.75 range or higher to be considered adequate. Individual company standards should be consulted.

Investor's (or Internal) Rate of Return

The investor's rate of return is the discount rate at which the sum of the discounted cash flows (excluding investments) equals the investment or at which the total present worth equals zero. The calculation procedure is a trial-and-error approach best done with a computer or calculator. For the example in Table C-1, the calculation would be as shown in Table C-2. The investor's rate of return in this example is 23%. Another form of rate of return is the "growth rate of return," in which the company's average reinvestment rate is included in the calculation. Ref. 1 provides an excellent discussion of this parameter.

Payout

Payout is the length of time required for the income from the project (discounted cash flow excluding investment) to recover the investment. Another way to put this is that payout is the length of time after the initial investment that it takes the cumulative discounted cash flow to equal zero. The weakness of this parameter is that it tells nothing about what happens after payout. This factor is calculated by summing the discounted cash flows by year (or month or other time period) until the total equals zero. For the example in Table C-1, the calculation for a 12% interest rate is shown in Table C-3. The payout discounted at 12% is 3 years and 3.3 months, or 3.27 years. Payout may also be calculated with cumulative, undiscounted cash flows. Obviously, this value will be less; i.e., the payout will be faster than the discounted method. Company policy will usually specify which is to be used in economic evaluations.

Profit-to-Investment Ratio

This is an undiscounted parameter that is simply the ratio of the total cash flow without investment divided by the total investment. The weakness of this parameter is that the timing of the investments and profits is not taken into account. It is a commonly used parameter but is not recommended. The profit-to-investment ratio of the example in Table C-1 would be calculated as profit/investment = (4,000+3,000+2,000+1,500+1,000+500)/(8,000). Therefore, the ratio would be 1.5 for this project.

Reference

1. Capen, E.C., Clapp, R.V., and Phelps, W.W.: "Growth Rate: A Rate-of-Return Measure of Investment Efficiency," *JPT* (May 1976) 531–43.

TABLE C-1—PRESENT-WORTH VALUE CALCULATION

Year	Cash Flow (dollars)	Discount Factor = 12%	Discounted Cash Flow
Initial investment	− 8,000	1.00	− 8,000
1988	4,000	0.94	3,760
1989	3,000	0.83	2,490
1990	2,000	0.74	1,480
1991	1,500	0.66	990
1992	1,000	0.58	580
1993	500	0.52	260
		Present-worth value = total =	1,560

TABLE C-2—INVESTOR'S RATE-OF-RETURN CALCULATION

Year	Cash Flow (dollars)	Discounted Cash Flow 15% Rate	Discounted Cash Flow 20% Rate	Discounted Cash Flow 23% Rate
Initial	− 8,000	− 8,000	− 8,000	− 8,000
1988	4,000	3,714	3,625	3,573
1989	3,000	2,398	2,226	2,129
1990	2,000	1,376	1,215	1,129
1991	1,500	888	746	672
1992	1,000	510	407	356
1993	500	219	167	141
Present-worth-value =		1,105	386	0

TABLE C-3—PAYOUT CALCULATION

Year	Discounted Cash Flow (dollars) Annual	Discounted Cash Flow (dollars) Cumulative
Initial	− 8,000	− 8,000
1988	3,760	− 4,240
1989	2,490	− 1,750
1990	1,480	− 270
First 3.3 months of 1991	270	0
Remainder of 1991	720	+ 720

Appendix D
State Regulatory Agencies

Alabama

Oil and Gas
State Oil and Gas Board
P.O. Drawer O
University, AL 35486
(205) 349-2852

*Underground Injection Control
(UIC)*
State Oil and Gas Board
As above.

Air
Air Pollution Control Commission
1751 Federal Dr.
Montgomery, AL 36109
(205) 271-7700

Alaska

Oil and Gas
Oil and Gas Conservation Commission
3001 Porcupine Dr.
Anchorage, AK 99501
(907) 279-1433

UIC
Oil and Gas Conservation Commission
As above.

Air
Air and Solid Waste Section
Dept. of Environmental Conservation
P.O. Box O
Juneau, AK 99811
(907) 465-2600

Arkansas

Oil and Gas
Oil and Gas Commission
314 East Oak St.
El Dorado, AR 71730
(501) 862-4965

UIC
Oil and Gas Commission
As above.

Air
Pollution Control and Ecology Dept.
P.O. Box 9583
Little Rock, AR 72219
(501) 562-7444

California

Oil and Gas
Div. of Oil and Gas
Dept. of Conservation
1416 Ninth St., Room 1310
Sacramento, CA 95814
(916) 445-9686

UIC
Div. of Oil and Gas
As above.

Air
California Air Resources Board
P.O. Box 2815
Sacramento, CA 95812
(916) 322-5840

Plus
Various area-wide and local air pollution control districts

Colorado

Oil and Gas
Oil and Gas Conservation Commission
1580 Logan St., Suite 380
Denver, CO 80203
(303) 866-3531

UIC
Oil and Gas Conservation Commission
As above.

Air
Air Pollution Control Div.
Colorado Dept. of Health
4210 East 11th Ave.
Denver, CO 80220
(303) 320-8333

Illinois

Oil and Gas
Div. of Oil and Gas
Dept. of Mines and Minerals
400 South Spring St.
Springfield, IL 62706
(217) 782-7756

UIC
Div. of Oil and Gas
Dept. of Mines and Minerals
Stratton Office Bldg., Room 704
Springfield, IL 62706
(217) 782-7756

Air
Illinois Environmental Protection Agency
Div. of Air Pollution Control
2200 Churchill Rd.
Springfield, IL 62706
(217) 782-2113

Indiana

Oil and Gas
Oil and Gas Div.
Dept. of Natural Resources
100 North Senate Ave.
Indianapolis, IN 46204
(317) 232-4055

UIC
U.S. EPA
Region V
230 South Dearborn Ave.
Chicago, IL 60604
(312) 353-2212

Air
Air Pollution Control Board
1330 West Michigan St., Room 436
Indianapolis, IN 46206
(317) 633-0611

Kansas

Oil and Gas
Oil and Gas Conservation Div.
Kansas Corporation Commission
200 Colorado Derby Bldg.
202 West First St.
Wichita, KS 67202-1286
(316) 263-3238

UIC
Kansas Corporation Commission
As above.

In cooperation with
Kansas Dept. of Health and Environment
State Office Bldg.
Topeka, KS 66612

Air
Bureau of Air Quality & Radiation Control
Kansas Dept. of Health and Environment
Forbes Field Bldg. 740
Topeka, KS 66620
(913) 862-9360

Kentucky

Oil and Gas
Oil and Gas Div.
Dept. of Mines and Minerals
P.O. Box 690
Lexington, KY 40586
(606) 257-3812

UIC
Oil and Gas Div.
As above.

Air
Div. of Air Pollution Control
Dept. for Environmental Protection
Fort Boone Plaza, 18 Reilly Rd.
Frankfort, KY 40601
(502) 564-3382

Louisiana

Oil and Gas
Louisiana Dept. of Natural Resources
Office of Conservation
P.O. Box 4275
Baton Rouge, LA 70804
(504) 342-5540

UIC
Louisiana Office of Conservation
UIC Section
As above.

Air
Louisiana Dept. of Environmental Quality
Air Quality Div.
625 N. Fourth St.
P.O. Box 44096
Baton Rouge, LA 70804
(504) 342-1206

Michigan

Oil and Gas
Dept. of Natural Resources
Geological Survey Div.
Stevens T. Mason Bldg.
P.O. Box 30028
Lansing, MI 48909
(517) 373-1837

UIC
Dept. of Natural Resources
As above.

Air
Dept. of Natural Resources
Air Quality Div.
As above.
(517) 373-7023

Mississippi

Oil and Gas
State Oil and Gas Board
P.O. Box 1332
Jackson, MS 39215
(601) 359-3725

UIC
State Oil and Gas Board
As above.

Air
Mississippi Air and Water Pollution Control Commission
P.O. Box 10385
Jackson, MS 39205
(601) 961-5171

Missouri

Oil and Gas
State Oil and Gas Council
Div. of Geology and Land Survey
P.O. Box 250
Rolla, MO 65401
(314) 364-1752

UIC
State Oil and Gas Council
As above.

Air
Air Pollution Control Program
Dept. of Natural Resources
P.O. Box 176
Jefferson City, MO 65101
(314) 751-4817

Montana

Oil and Gas
Board of Oil and Gas Conservation
1520 Sixth Ave.
Helena, MT 59620
(406) 444-6675

UIC
Region VIII, U.S. EPA
999 18th St., Suite 1300
Denver, CO 80202

Plus
Board of Oil and Gas Conservation
As above.

Air
Environmental Sciences Div.
Dept. of Health and Environmental Sciences
Cogswell Bldg.
Helena, MT 59620
(406) 444-3454

Nebraska

Oil and Gas
Oil and Gas Conservation Commission
P.O. Box 399
Sidney, NE 69162
(308) 254-4595

UIC
Oil and Gas Conservation Commission
As above.

Air
Air Quality Div.
Dept. of Environmental Control
P.O. Box 94877
Lincoln, NE 68509
(402) 471-2186

New Mexico

Oil and Gas
Oil Conservation Div.
Energy, Minerals, and Natural Resources Dept.
P.O. Box 2088
Santa Fe, NM 87501
(505) 827-5800

UIC
Oil Conservation Div.
As above.

Air
Environmental Improvement Agency
Air Quality Bureau
P.O. Box 968
Santa Fe, NM 87504-0968
(505) 827-0070

North Dakota

Oil and Gas
Oil and Gas Div.
North Dakota Industrial Commission
900 East Blvd.
Bismarck, ND 58505
(701) 224-2969

UIC
Oil and Gas Div.
As above.

Air
Environmental Engineering Div.
Health Dept.
1200 Missouri Ave., Room 304
P.O. Box 5520
Bismarck, ND 58502
(701) 224-2348

Ohio

Oil and Gas
Div. of Oil and Gas
Ohio Dept. of Natural Resources
Fountain Square, Bldg. A
Columbus, OH 43224
(614) 265-6917

UIC
Div. of Oil and Gas
As above.

Air
Ohio Environmental Protection Agency
P.O. Box 1049
Columbus, OH 43215
(614) 466-8565

Oklahoma

Oil and Gas
Oklahoma Corporation Commission
Oil and Gas Div.
Jim Thorpe Bldg.
Oklahoma City, OK 73105
(405) 521-2302

UIC
Oklahoma Corporation Commission
Underground Injection Dept.
As above.

Air
Air Quality Service
State Dept. of Health
1000 NE 10th
P.O. Box 53551
Oklahoma City, OK 73152
(405) 271-5220

Pennsylvania

Oil and Gas
Bureau of Oil and Gas Management
Dept. of Environmental Resources
P.O. Box 2057
Harrisburg, PA 17120
(717) 783-9645

UIC
Region III, U.S. EPA
6th and Walnut Streets
Philadelphia, PA 19106

Plus
Bureau of Oil and Gas Management
As above.

Air
Air Quality Control Bureau
Dept. of Environmental Resources
18th Floor Fulton Bldg., 200 N. 3rd St.
P.O. Box 2063
Harrisburg, PA 17120
(717) 787-9702

Texas

Oil and Gas
Oil and Gas Div.
Texas Railroad Commission
P.O. Drawer 12967, Capitol Station
Austin, TX 78711
(512) 463-6887

UIC
Texas Railroad Commission
Director of Underground Injection Control
As above.

Air
Texas Air Control Board
630 Highway 290 East
Austin, TX 78723
(512) 451-5711

Utah

Oil and Gas
Div. of Oil, Gas, and Mining
355 W. North Temple
3 Triad Center, Suite 350
Salt Lake City, UT 84180-1203
(801) 538-5340

UIC
Div. of Oil, Gas, and Mining
As above.

Air
Air Quality, Dept. of Health
P.O. Box 16690
Salt Lake City, UT 84116-0690
(801) 538-6108

West Virginia

Oil and Gas
Oil and Gas Conservation Commission
1613 Washington St., East
Charleston, WV 25311
(304) 348-3092

UIC
Oil and Gas Conservation Commission
As above.

Air
Air Pollution Control Commission
1558 Washington St., East
Charleston, WV 25311
(304) 348-3286

Wyoming

Oil and Gas
Oil and Gas Conservation Commission
P.O. Box 2640
Casper, WY 82602
(307) 234-7147

UIC
Oil and Gas Conservation Commission
As above.

Air
Div. of Air Quality
Dept. of Environmental Quality
Herschler Bldg.
122 West 25th St.
Cheyenne, WY 82002
(307) 777-7391

Author Index

A

Abel, W.L., 59
Adams, G.H., 15, 23
Al-Arfai, K.A., 23
Althouse, W.S., 52
American Iron and Steel Inst., 52
American Natl. Standard Inst., N-45.2.10, 74, 75
American Petroleum Institute (API),
  Manual on Disposal of Refinery Wastes, 23
  RP-10E, 49, 52
  RP-38, 23
  RP-45, 23
  Standard 7B-11C, 29, 45
  Standard 610, 73, 75
  Standard 674, 73, 75
American Soc. of Mechanical Engineers, 79
Amezcua, J.D., 54, 59
Arco Chemical Co., 24
Arco Oil & Gas Co., 52
Arnold, K.E., 73–75

B

Baer, P.J. Jr., 10
Baker, S., 19
Barkman, J.H., 15, 23
Bartos, W.B., 49, 52
Battle, J.L., 15, 23
Behrenbruch, P., 64
Bertrem, B., 59
Bilhartz, H.L., 13, 23
Blanton, M.L., 45
Bowman, R.W., 17, 23
Bowyer, P.M., 23
Bradley, B.W., 23
Brigham, W.E., 10
Brown, K.E., 54
Bryan, R.W. and Co., 71
Buckhouse, A.R., 64
Buckwalter, J.F., 25, 45
Burchard, H.G., 52
Byars, H.G, 21, 23

C

Calberg, B.L., 24
Campbell, E.E., 52
Capen, E.C., 81
Carlberg, B.L., 23
Case, C.A., 10
Castagno, J.L., 23
Caterpillar Tractor Co., 45
Caudle, D.D., 23
Cerini, W.F., 15, 23
Chen, E.Y., 19, 23
Chen, R.B., 19, 23
Cheng, K.J., 23
Chilton's Instrument and Control Systems, 23
Clapp, R.V., 81
Clawson, R.M., 75
Clementz, D.M., 23
Cloughley, T.M.G., 75
Coffey, L.F., 64
Costerton, J.W., 23
Cox, E., 59
Craft, W.S., 75
Craig, F.F. Jr., 1, 2, 4, 5, 7, 10, 25, 26, 45
Crane Co., TP410, 52

D

Davidson, D.H., 15, 23
Davidson, L.B., 64
Davis, L.E., 23

Degner, V.R., 23
Diesel Engine Manufacturers Assn. (DEMA), 29, 45
Doscher, J.M., 23
Dozier, H.S., 56, 59

E

Dymond, P.F., 53, 54
Eager, R.G., 24
Einstein, A., 59
Emerick, G.M., 45
EnDean, H.J., 52
Evers, J.F., 64
Evers, R.H., 19, 23

F

Fast, C.R., 26, 45
Finch, E.M., 23
Fowler, E.D., 52
Frank, W.J., 23
Frankhouser, H.S., 74, 75
Fraylick, J.R., 75
Fulks, B.D., 2, 4
Funk, R.J., 52

G

Gallop, B.R., 23
Garibaldi, C.A., 64
Gas Processors Assn., 45
Gates, G.L., 23
Geesey, G.G., 23
Ghauri, W.K., 53, 54, 59
Gleeson, C.W., 37, 45–47, 52
Gossett, J., 59
Greenhorn, R.A., 10
Gregg, D.E., 72, 75

H

Haberthur, C.R., 10, 45, 52, 71
Hammerlindl, D.J., 52
Hardy, I.J., 59
Harrison, O.R., 56, 59
Hensel, W.M. Jr., 54
Henshaw, T.L., 38, 45
Hewitt, C.H., 14, 23
Hill, V.F., 72, 75
Holland, A.L. Jr., 40, 45
Howard, G.C., 26, 45

I

Iskander, F.F., 64

J

Jackson, R.W., 10
Johnson, C.R., 10
Johnson, J.L., 40, 45
Johnson, K.H., 23
Jones, F.O. Jr., 23
Jones, R., 64
Jorque, M.A., 23
Jowell, J.H., 10, 45, 52, 71
Junkins, E.D. Jr., 52

K

Kamal, M., 10
Karassik, I., 39, 45
Kendall, R.F., 67
Kennedy, J.L., 22, 23
King, P.J., 23
Kissel, C.L., 23

Kohan, D., 52, 59
Kostrop, J.E., 45

L

Langewis, C. Jr., 37, 45–47, 52
Lea, J.F., 59
Leech, C.A. III, 23
Lockyer, K.G., 75
Logan, J.L., 52
Lubinski, A., 52

M

Maharidge, R.L., 23
Marquis, D.E., 32, 33, 45
Mathur, A., 47, 52
Maxwell, R.L., 64
Maxwell, W.N., 54
McCume, C.C., 15, 23
McDonald, J.P. Jr., 23
McDuff, J.R., 21, 23
McFarlane, R.C., 64
McGraw-Hill Book Co., 45
Megill, R.E., 62, 64
Meyer, W.L., 75
Miller, J.E., 36, 37, 45
Mitchell, R.W., 23
Moe, R.C., 31, 45
Moore, B.L. Jr., 45, 52, 59
Moore, H.J., 45
Moran, J.P., 54
Morgen, R.A., 56, 59
Mueller, T.D., 64
Muther, R., 2, 4

N

Natl. Assn. of Corrosion Engineers (NACE):
  RP-0-69, 24
  RP-02-78, 23
  RP-03-72, 24
  RP-04-75, 24, 40, 45, 52
  RP-05-75, 24
  RP-07-75, 23
  TM-01-73, 23
Neal, J.C., 45
Nelson, K.C., 54, 59
Nomitsu, T., 23

O

O'Keefe, W., 59
Ostroff, A.G., 13, 14, 23

P

Pace, F.A., 59
Parent, C.F., 23
Patton, C.C., 15, 23
Perry, L.N., 23
Phelps, W.W., 81
Pipeline Ind., 45, 59
Pratos, C.A., 21, 23
Preuninger, P.D., 45
Pumphrey, D.M. Jr., 52

R

Reed, M.G., 23
Reynolds, F.S., 64
Rhodes, A.F., 52
Ritter, J.B., 75
Robinson, K., 23
Roye, J.E. Jr., 52, 59
Ruseska, I., 23

Subject Index

www.ingramcontent.com/pod-product-compliance
Lightning Source LLC
Chambersburg PA
CBHW081551220326

41598CB00036B/6638